Interior Decorating: Paints and Wallcoverings

New Illustrated Library of Home Improvement Volume 2

Interior Decorating: Paints and Wallcoverings

Prentice-Hall/Reston Editorial Staff

Prentice-Hall of Canada, Ltd. / Reston Publishing Company,
Scarborough, Ontario.

Series contributors/ H. Fred Dale, Richard Demske, George R.
Drake, Byron W. Maguire, L. Donald Meyers, Gershon
Wheeler

Design/ Peter Maher & Associates
Color photographs/ Peter Paterson/Photo Design

Printed and bound in Canada.

The publishers wish to thank the following companies for
providing photographs for this volume:

Academy Handprints, Ltd.
Canadian Wallpaper Manufacturers, Ltd.
Columbus Coated Fabrics
F. Schumacher and Co.
Georgia-Pacific Corporation
Masonite Corporation
PPG Industries
Richard E. Thibaut, Inc.
Standard Coated Products, Inc.
Wallcovering Industry Bureau
Wooster Brush Company

Contents

Interior Decorating: Paints and Wallcoverings

The Need for Paint

A coat of paint quickly brightens a dull room or a drab house. The improvement is immediately noticeable, even before the job is finished. Paint makes a home more attractive inside and out and thus creates a pleasanter environment. From a purely monetary standpoint, a relatively inexpensive coat of paint adds value to the home far above the cost of the job.

The psychological value of paint is also important. Proper choice of colors can make a room seem warm or cool, exciting or comfortable, or larger or smaller than it actually is.

The most important reason for painting, however, is protection. Wood and other building materials are affected by the elements, are attacked by insects and small animals, become corroded from moisture, and are subject to damage from man-made pollutants in the atmosphere. Paint is the most economical protection against all these dangers.

The most obvious use for paint is as protection from the elements. Sun, rain, snow and wind all cause deterioration which can be prevented by a good coat of paint. The heat of the sun will dry out untreated wood, so that the wood becomes brittle and cracks. Paint on the surface acts as an insulating layer, absorbing or reflecting the sun's energy, and thus protecting the wood from drying. Rain and snow deposit moisture on the surface where it is absorbed into the material, causing wood to rot or warp and iron to rust. A good coat of paint on the surface keeps the moisture out of the building material. Wind whips up dust particles and blows them against a house, causing abrasion which wears away material. Paint is a protective coating which is able to withstand this abrasion and which can be replaced in any case when it does wear away. Note that paint on the surface protects the whole structure, since most deterioration starts at the surface.

Heat and cold cause another problem. Materials generally expand with heat and shrink with cold, and changes in temperature can loosen nails or cause splits in materials. Paint helps maintain a more nearly constant temperature in the interior and thus minimizes the chances of damage.

Another source of damage to a home is man-made pollution. Chemicals in smoke and gases from auto exhausts and industry can cause serious damage to wood, metal and other building materials. These pollutants are found almost everywhere in today's industrial society. Fortunately a coat of paint is sufficient protection to keep them away from the surfaces of buildings.

Inside the house, paint is also important as a protective coating. In bathrooms where water vapor from baths and showers condenses on walls and ceilings, an unpainted surface would be damaged quickly. Too much moisture can warp wood or cause plaster walls and ceilings to disintegrate. In kitchens, paint protects against moisture, grease and cooking vapors which attack walls and ceilings in much the same way pollutants do.

Paint is also important to protect the inside of a home from normal wear. Paint or varnish on a floor prolongs the life of the surface. Without the coating, floors would wear quickly from ordinary traffic. Paint protects walls from greasy hands and scrapes from children's toys. When the paint shows signs of wear, it can be replaced more easily and more cheaply than a section of floor or wall.

Special paints are sometimes used for extra protection against specific dangers. Thus, fungicides and mildew preventatives are added to exterior paints. Fireproof paint is used where there is a risk of fire. Although the material under the paint can burn, chemicals in the paint release gases when heated, and these form an insulating layer and tend to smother the flames.

In a strict sense, the word "paint" refers to an opaque coating applied to a surface. However, in general usage any material that is applied with painting tools is commonly called paint. This includes penetrating stains as well as transparent surface coatings such as varnish and lacquer. All these types of "paint" are discussed in the next chapter.

Paints

The chemistry of paint is a continuously developing science: paints today bear little resemblance to those used a decade ago. Although the specific ingredients change, there are four basic types of materials that go into paint. These are pigment, vehicle, thinner and drier. Most paints have all four elements, although in some a single chemical may serve two functions. Transparent lacquer or varnish has no pigment, and in fact many paints lack one or more of the four elements.

2-1. Pigments

Pigment is the solid material that gives paint its color and hiding ability; it is left on the surface as a hard coating when the paint has dried. Originally pigments were natural substances — usually minerals, but occasionally derived from plants and animals. Most pigments today, however, are synthetic. The pigment is always ground to powder before it is mixed in the paint.

The *coverage* of a paint is the area that can be covered with a given quantity, usually a gallon. Coverage is a function of the amount of pigment in the paint — generally, the more pigment, the greater the area that can be covered.

White pigments are usually used as a base in both white and colored paints. For colors other than white, a small amount of colored pigment is added. Originally, the most common and desirable pigment was *white lead.* White lead paint consisted of white lead pigment in an oil vehicle. It was very popular until it was discovered that white lead is poisonous, and it quickly dropped from favor. A small amount of white lead, not enough to be dangerous, is still used with other pigments in some paints.

The most economical white pigment is *lithopone,* a synthetic material. It is quite satisfactory for interior painting, but cannot withstand the elements as well as some other pigments. It is never used as the main pigment in outdoor paints, but it is sometimes added in small quantities to other pigments in exterior paints to lower the cost.

Titanium oxide is a pigment that is a product of the space age. Titanium was a laboratory curiosity until uses were found for it in space flights and electronics. As a pigment, titanium oxide has much better hiding ability than white lead. However, since titanium oxide is more expensive, it is usually blended with other white pigments and other materials. For example, in exterior paints, calcium carbonate is mixed with titanium oxide. As the paint ages, calcium carbonate forms a chalk-like dust on the painted surface. This "chalk" washes off in the rain, removing dirt with it, so that the surface appears clean and bright for the life of the paint job. Titanium oxide can also be combined with small amounts of white lead and lithopone.

Some colored pigments are obtained from the earth: for example, from red clay. However, most colored pigments used today are synthetic. By controlling the manufacturing process, paint manufacturers can reproduce exactly the man-made colored pigments, so that one batch of paint will be the same color as another. This is not always true of natural pigments.

2-2. Extenders

Extenders are other solid materials in paints. Because they are solid, they are usually included in the broad class of pigments; but they have pratically no hiding ability and do not increase the coverage of the paint, as true pigment would. Extenders do have specific uses, however, and are not simply fillers.

One common extender is ordinary *sand,* or *silica.* When paint dries with a hard slick surface, a second coat will not adhere to it readily. It would be possible to roughen the first coat by rubbing it with sandpaper, but a simpler method is to mix a small amount of sand with the first coat, as an extender. When the paint dries, the sand makes the surface rough enough to enable the next coat to adhere better. This roughness is called *tooth,* and a small amount of tooth is desirable.

Too much sand in the paint makes it heavy and reduces its coverage. When white lead paint was popular, painters frequently judged a paint by its weight. The heavier the paint was, the more white lead it presumably contained, and therefore the greater would be its coverage. Unfortunately, unscrupulous paint dealers sometimes added excessive sand to make the paint heavy, since sand was much cheaper than white lead. As with all extenders, a small amount is useful, but too much spoils the paint. The percentage of extenders in the paint is indicated on the label, but if you don't know the proper limits, it is best to rely on reputable paint manufactur-

ers. Bargain paints are usually no bargain.

Calcium carbonate, commonly called *whiting,* is another extender that causes chalking of exterior paints, as described in the preceding pages. Again, only a small amount is necessary. Too much whiting will cause too rapid chalking, so that the paint job looks streaked and wears out too rapidly. Chalking is desirable, but should not begin for a year or two after the paint is applied. If it starts too soon, it is a sign that too much whiting was added to the paint.

Talc and *alumina* are very light powders that are sometimes added to paints to help keep the heavier pigments in suspension. Other extenders prevent or reduce such paint defects as cracking and flaking.

2-3. Thinners

A liquid called *thinner* or *solvent* is sometimes added to paint to make it easier to apply. The thinner, which evaporates as the paint dries, is usually added to the paint mixture in manufacture, but for special methods of application it may be necessary to add additional thinner. Oil-base paints use turpentine or mineral spirits as a thinner, and mineral spirits are also used to thin alkyd paints. Water-base paints are thinned with water. Kerosene, benzine and other flammable thinners should be used with caution and stored carefully.

Thinner may be used to clean brushes, rollers and other tools after a paint job is finished. Washing the painting tools in soapy water will remove the thinner.

2-4. Driers

Paints dry in three different ways, and a specific paint may use more than one method. Anything added to the paint to speed drying is called a drier.

Evaporation is one process of drying, and to some extent all paints dry by evaporation. However, some paints dry only by this method. Water-base paints dry by the evaporation of water. Shellac dries by the evaporation of the alcohol it contains. A paint that contains any volatile thinner also dries by evaporation.

A second drying process is *oxidation*. The oil in oil-base paints unites chemically with oxygen in the air to form a hard film. For this reason oil in paint is sometimes called *drying oil*. Oil added to alkyd paints speeds drying in this fashion.

The third drying process is *polymerization*, a chemical combining of liquids to form a solid. Epoxy paints and polyurethane coatings dry in this way. The resultant coating is very hard and tough.

2-5. Vehicles

If the pigment used in paint were just rubbed on a surface, it could not be applied evenly and would not adhere well. To facilitate spreading the paint easily and evenly, the pigments are ground into fine powder and stirred into a liquid called a *vehicle* or *carrier*. The pigments do not dissolve in the liquid, but remain suspended as in an emulsion. The vehicle may be a separate liquid from the thinner and drier used in the paint; or one liquid, such as water, may serve two or even all three functions. For this reason it is customary to refer to the liquid part of paint as the vehicle, even though it may be a blend of vehicle, drier and thinner.

At one time, oil was the only vehicle used. Oil is also a drier, as described in the preceding pages. Linseed oil was preferred, but other vegetable and animal oils have been used, including castor oil, soybean oil and fish oil. In modern oil-base paints, synthetic resins add strength to the paint, making the coating hard and tough.

Alkyd is a synthetic resin formed by combining an alcohol and an acid. Alkyd is combined with a drying oil, and the combination dissolved in a thinner to make a liquid which is used as the vehicle in alkyd paints.

Latex paints use a water emulsion as the vehicle (an emulsion is a mixture in which tiny particles are held in suspension in liquid). The emulsion in latex paint is a suspension of a synthetic resin in water. The first latex paint used butadiene-styrene (BDS) as a resin. Since BDS is a kind of artificial rubber called latex, the paint was called latex paint, and later other paints using any resin emulsion in water were also called latex. Some of the resins used are polyvinyl acetate (PVA), polyester and acrylic. Latex paint can be thinned by adding water, and water is all that is needed to clean up afterwards.

2-6. Types of Paint

When oil was the principal vehicle, paints were usually designated by the type of pigment they used: e.g., *white lead* paint. As different vehicles were developed, the characteristics of a paint depended more on the vehicle than the pigment. Thus an oil-base paint with lithopone and an oil-base paint with titanium oxide have more in common than an oil-base paint and a latex paint which both use lithopone as a pigment. Consequently, paints are usually designated by their vehicles, even though there may be no trace of the vehicle after the paint dries. The three principal vehicles in paint are oil, alkyd and latex.

Paint is also designated by its purpose. House paint is used on exteriors and is specially formulated to withstand the elements. Floor paint is a tough coating which withstands the abrasive action of foot traffic on floors and decks. *Marine paint* is an expensive, high-quality paint used on boats and formulated to withstand the effects of salt water and weather. *Masonry paint*, as its name implies, is designed to adhere better to masonry than ordinary paints do.

In addition to being named for its vehicle and purpose, paint may also be designated by its finish. *Flat paint* has a flat, lusterless finish. *Glossy paint* is shiny. The luster is achieved by adding special oils that dry to a shiny film on the surface. Glossy paints may be further separated into *high-gloss* paints and *semi-gloss* paints, depending on the degree of sheen. *Primers* or *primer-sealers* are used as a first coat on surfaces that will not take paint well. The primer is a material that will adhere to the surface, often sealing it, if it is porous, and will readily take a coat of regular paint. *Enamel* is a very hard finish used where resistance to abrasion, wear or moisture is needed. Note that enamel and gloss paints are two different materials. Enamels usually have a glossy finish, but they are also available in flats.

A complete designation of a paint might include the vehicle, the purpose and the finish. For example, *alkyd semi-gloss wall paint* is a complete designation. Latex house paint does not indicate the finish, but house paints are assumed to be flat unless otherwise specified. *Alkyd floor enamel* is an alkyd-base enamel and since it is used on floors, it is usually glossy.

2-7. Oil-base Paints

At one time oil-base paints were the preferred paints and, in fact, almost the only paints used by professional painters. Oil-base house paint is still used occasionally for exteriors, but for interiors it has been superseded by alkyd and latex paints.

Oil-base paints are available in flat, semi-gloss and gloss finishes. This paint takes two to four days to dry and has a strong odor which persists even after the paint is completely dry. The paint dries by oxidation of the oil. The paint can be applied with a brush, roller or other applicator, or a spray gun.

Oil-base paints have better hiding ability and greater coverage than latex. Also, oil-base paints can be applied to bare wood, to metal and to plastic. These paints cannot be applied to damp walls.

Oil-base paints can be thinned with mineral spirits or turpentine. Surfaces to be painted can also be cleaned with these materials, and after the job is done, tools can be cleaned in the same thinner. Kerosene is a cheap solvent for oil-base paints and can be used for clean-up in place of mineral spirits or turpentine. Kerosene should not be used as a thinner in the paint itself. It is flammable and storage may be a problem.

Oil-base paint may still be available in paint stores, but it is rapidly becoming obsolete, even for outdoor work. Alkyd paints or latex paints are preferred for every type of paint job.

2-8. Alkyd Paints

The development of alkyd paints revolutionized the paint industry because alkyd paints are superior to oil-base paints in so many respects. Some of the advantages of alkyds over oil-base paints are as follows:

1. Alkyd paints dry in six to twenty-four hours compared to two to four days for oil-base paints.

2. Alkyds are tougher and can stand more scrubbing and abrasion.

3. Alkyds have very little objectionable odor, as oil-base paints do.

4. Alkyd paints last longer than oil-base paints.

5. The coverage of alkyd paint is excellent and on porous masonry or dried wood it is better than that of oil-base paints.

6. Alkyd paints produce a tight paint film that is water-resistant. For bathroom walls and other areas where there may be moisture, alkyds are preferred.

WARNING: Although alkyd paints are odorless, their fumes are toxic and flammable until the paint dries. Adequate ventilation must be provided when painting with either alkyd or oil-base paint.

Alkyd paints do have an oil base and can be thinned with mineral spirits just as oil-base paints can. Turpentine could also be used, but turpentine has a strong odor and should not be used in an "odorless" paint. Either turpentine or mineral spirits can be used to clean up after the job is finished. Kerosene may also be used to clean the paint tools, but it is better to avoid flammable solvents.

Alkyd paints are available in flat, semi-gloss and gloss finishes. So-called alkyd enamel may be high-gloss alkyd paint and not a true enamel. On interior woodwork, alkyd paints provide a tougher finish than latex. Some manufacturers provide both alkyd and latex paints in matching colors so that latex can be used on the walls and alkyd on the woodwork. This makes an outstanding paint job, but also involves more work, since different thinners are used for the two paints, and two different methods of cleaning are necessary when the job is finished.

Alkyd paints are easy to apply by brush, roller or spray gun. They do not show lap marks. These paints can be used directly on most surfaces without a primer. They cannot be applied to damp surfaces, however. Alkyds dry by oxidation of the drying oils in the paint.

2-9. Latex Paints

The development of latex paints was a huge step forward in the paint revolution that started with the introduction of alkyd paints. Latex threatens to make all other paints obsolete. Latex paint uses a vehicle consisting of an emulsion of a synthetic resin in water, and because water is used as the solvent, latex paint has many advantages, some of which are as follows:

1. It dries (by evaporation of water) in an hour or two, much faster than alkyds do.

2. It is applied easily by brush, roller or spray gun without lap marks. Painting with latex takes less effort than with other paints.

3. Cleaning up after the job is finished is simple, since tools can be cleaned in water rather than special thinners.

4. It is odorless, and there are no dangerous fumes while it is drying.

5. Colors stay bright for years.

6. It can be applied on damp surfaces, since water is part of the paint anyway. This means you can paint a house when there is morning dew on the wall or even when the plaster is new.

7. Porous coating permits moisture below to evaporate. This prevents peeling and blistering.

8. It resists alkalis which can damage alkyd paints. This makes latex excellent for masonry, which frequently contains alkali.

9. It can be applied directly to metals.

10. It withstands mildew.

11. It dries to a flexible surface that stretches or bends with movement of the surface below it.

Latex paint does have a few disadvantages:

1. It cannot be applied directly to new wood without a primer.

2. It does not adhere well to high-gloss oil-base finishes.

3. It does not withstand scrubbing as well as alkyds do.

Despite its disadvantages, latex paint is by far the "one best paint" for both interiors and exteriors. Development of latex paints is continuing, and the disadvantages are being minimized or eliminated.

Latex paint is available in flat, semi-gloss and gloss finishes. There is also a *dripless* variety with a much thicker consistency that is supposed to be advantageous in painting ceilings. Even with dripless paint, however, it is impossible to paint a ceiling without dripping, and the thicker paint is more expensive and requires more effort to apply than ordinary latex.

2-10. Water-thinned Paints

Latex paints are water-thinned, but the term *water-thinned paint* refers to materials that are not emulsions, including whitewash, calcimine and casein paints. These are not really paints, but washes. They are not durable and not easily washed without scrubbing off part of the coating. Water-thinned paints were used because they were inexpensive and easy to apply, and because they dried quickly. They are seldom used now except where a quick inexpensive coating is needed.

Whitewash is a mixture of lime and water. *Calcimine* is a mixture of chalk and glue. It is sold in powdered form, and is mixed with water for use. *Casein paint* comes from skim milk. It is also sold as a powder to be mixed with water as needed.

Portland cement paint is a special water-thinned mixture used on masonry. It is sold as a powder to be mixed with water. It is not really a paint, since it doesn't dry as a paint, but rather sets as concrete does. It is applied to damp masonry and must be kept moist for two or three days before it is allowed to harden. It is cheap and forms a surface that will take any kind of paint. Colored pigments are sometimes added to portland cement paint.

2-11. Rubber-base Paints

Rubber-base paints, as their name indicates, use a synthetic rubber as a base. The rubber is thinned in a volatile solvent such as naphtha or turpentine. These paints are tricky to apply since they dry very quickly. They were originally designed for use on concrete, stucco, brick and other masonry surfaces and are most effective for abrasion resistance and waterproofing. Rubber-base paints require special solvents for cleaning up. For masonry, latex masonry paint has largely replaced these paints, but rubber-base paint is still used for painting swimming pools.

2-12. Primers, Sealers and Undercoaters

Surfaces that have never been painted before, such as bare wood, gypsum board, or metal, usually require more than one coat of paint. The first coat of paint is called *primer,* and all the coats together form what is called a *paint system.* The primer is usually a different kind of paint from the top coat if the top coat is a paint that will not adhere well to the untreated surface. The primer must be compatible with both the surface to be painted and the top layer of paint. That is, the primer must adhere well to the surface, and when it has dried, the top coat should adhere well to the primer. The primer must have tooth or roughness so that the finish coat can stick to it. In general, paint will not adhere to a surface that is too slick.

A primer is also used when the top coat does not furnish satisfactory protection for the surface to be painted. For example, metal can rust even under a coating of paint. A protective primer to prevent rust is usually used on metal before the final coat of paint is applied.

Sometimes defects or chemicals in a surface can mar a paint job. For example, resin from knotholes can "bleed through" the paint weeks or months after the finish is dry. In this case, a first coat of a special sealer is used to seal in any chemicals that can damage the finish. This is technically called a *primer-sealer* since it is a first coat. Sealers are also used to "seal" porous materials so that they do not absorb the moisture in the paint, which could cause imperfect drying.

Primer-sealers are used on bare wood which is to be coated with a clear finish. Varnish, lacquer and stains are absorbed into bare wood and raise a nap which is rough to the touch. Sealers for this purpose are clear so that the natural grain of the wood will be visible.

When a new coat of paint is to be applied to a surface which has once been painted satisfactorily, no primer is required. Of course, if the old paint job was poorly done, or especially if paint is peeling and exposing bare wood, it is best to proceed as though the surface had never been painted and to use a primer. However, a surface which has once been painted has been sealed, if not by a primer, then by the paint itself. It may still be necessary to apply two coats, especially if a dark wall is to be covered with a light paint. In this case, the first coat applied over the old paint is called an *undercoat,* and need not be of as high a quality as the finishing coat. When the finish coat is enamel or a gloss paint, the undercoat should be a flat paint since the finish coat will stick better to a flat than to a sleek coating.

Latex should never be applied directly to bare wood. For the best job, an alkyd primer should be used. However, if you use alkyd for one coat and latex for another, you have two different cleanup jobs. It is not too bad if the alkyd paint is applied at one time and the latex at another, but it becomes a problem if you wish to put a primer on bare wood and at the same time put a latex coating somewhere else. The solution is to use a special latex primer that is available for exteriors. It is somewhat different from ordinary latex paint in that it can be used on bare wood, and it takes about eight hours to dry, instead of the hour or two for conventional latex. However, it is thinned with water, and water is used for cleaning up, as with other latex paints.

Similar considerations apply when the top coat is to be an alkyd paint. For some applications, latex may be a better primer, but for simplicity in cleaning up (this is said to be the worst part of a paint job), try to use one kind of paint for all coats. For example, latex is better for the first coat on masonry because it is resistant to damage from alkalis. Again, there are alkyd primers available for covering masonry.

Metal that can rust should be painted with a rust-inhibiting primer before any other paint is applied. The primer used on metals must adhere well to the metal and also must ensure a good bond to the finish coat. Zinc chromate is an excellent primer for all metal since it prevents rust and also provides a coating that will take any other paint with a good bond. However, zinc chromate is not available in a water-thinned vehicle. There are latex metal primers available, which, though not as effective as zinc chromate, do have the advantages that go with latex paints. Outdoors, all metal surfaces should have a primer applied before the finish coat. Inside the house, metal surfaces are less apt to rust and therefore need be primed only if the paint to be used will not stick well. Latex paint can be applied to bare metal inside the house, and usually one coat is sufficient.

Plaster and gypsum board will not take alkyd paints, but latex adheres nicely to either. However, because of the porosity of both of these materials and the likelihood of raising the nap on gypsum board, the first coat is more difficult to apply and gives less coverage than subsequent coats. The best primer is a latex primer-sealer which is really only a thin latex paint. Because of its thinness, it goes on easily and seals the pores in the material. One coat of ordinary latex paint over the primer completes the job. Like ordinary latex, this interior latex primer dries quickly.

2-13. Clear Finishes

Clear finishes contain no pigment and thus have no "hiding" ability as paints do. They are used generally to finish surfaces so that the natural grain is visible. They are sometimes called *natural finishes.* Although some of these natural finishes are hard and tough enough to withstand abrasion that would quickly wear paint, they do not wear as well when exposed to sun and extremes of heat and cold. The pigment in paint is a protection against the elements which is missing in clear finishes.

Shellac is a clear finish made by dissolving an insect resin, lac, in alcohol. It dries in about fifteen minutes and forms a hard coating which can seal plaster or gypsum board and prevent resins, alkalis and other impurities from bleeding through the surface. For these reasons, shellac is an excellent primer-sealer for most paints. It is also used as a clear finish on floors and furniture, but must be protected with a wax coating. Shellac can be applied by brush or spray gun.

Shellac is usually packaged in a *4-pound* or *5-pound cut.* The cut refers to the amount of resin dissolved in one gallon of alcohol. It should always be thinned to something between a 2-pound and 3-pound cut before using. Ask your paint dealer how much alcohol to add to the cut you buy to get a 2-1/2 – pound cut. The exact cut doesn't matter, but two thin coats of shellac give a longer lasting finish than one thick one. For small quantities, the dealer will usually sell you already-thinned shellac.

Old shellac does not dry well. Look for a date on the package and try not to buy shellac that is more than 6 months old. If the can is undated, ask to try a small amount on a scrap of wood. It should get tacky in a few minutes. Because shellac ages quickly, don't buy more than you need at one time.

Shellac is available in two types: orange and white. Orange is cheaper and is a good primer under opaque finishes. White shellac should be used when a clear finish is desired. Since shellac is thinned with alcohol, dena-tured alcohol is also used for cleaning brushes and other equipment, when the job is finished. For this reason, shellac should not be used to finish tables or bars where alco-holic drinks may be spilled. A shellac finish also gets cloudy when water is spilled on it. However, on furniture that is not subject to spilled beverages, shellac makes a nice looking, inexpensive finish.

Lacquer is a clear finish that dries very rapidly and can be applied only with a spray gun. Special slow-drying lacquers are availa-ble for brushing but they are inferior to the fast-drying variety. Lacquer consists of nitro-cellulose dissolved in a volatile solvent and is thinned with a special mixture of solvents called *lacquer-thinner.* The same mixture is used to clean the spray gun after the job is done. Since lacquer acts as a paint remover, it cannot be applied over old paint coatings, but it can be used over shellac. Lacquer is difficult to apply, and considerable practice is required before it can be put down correctly. It is also highly flammable. For these reasons, home handymen should probably avoid using lacquer.

Varnish is a mixture of resin, oil, thinner and dryer. However, the proportions and types of ingredients can be varied to produce the best varnish for a particular purpose. No one varnish can be used on everything. A stiff brush should be used to apply varnish. The brush should not be used for anything else, since even a small amount of pigment from another job can spoil a varnish finish. Varnish can also be applied with a lint-free cloth pad. Varnish must not be stirred, since stirring causes bubbles which cannot be brushed out. In fact, the formation of bubbles is the only problem that arises when varnish is applied. Tools can be cleaned in turpentine or mineral spirits.

When varnish dries, first the solvent evapo-rates, then the drying oils oxidize. Drying normally takes about twenty-four hours. The dry coating is clear, hard and long-lasting. When composed of the proper synthetic resins and oils, the coating can withstand water, alcohol, most solvents, high or low temperatures and severe abrasion.

Floor varnish is formulated to withstand the heavy abrasion of foot traffic and scraping of furniture across the floor. It can be scrubbed clean with soap and water and can be waxed. No other varnish produces as hard a finish. Floor varnish dries more rapidly than other varieties, usually taking twelve to sixteen hours.

Cabinet finishing varnish is a clear material that dries to a hard coat and can be polished to a high gloss. It dries in twenty-four hours.

Spar varnish, a very expensive variety designed for exterior and marine use, will withstand the elements and wide variations in temperature. It is not used inside the house, however, because it cannot stand abrasion as well as floor varnish can and cannot be polished to as fine a sheen as cabinet finishing varnish.

There are many other types of varnish available in finishes from a dull flat to an extra-high gloss. The drying time, too, varies from two hours to two days. An old rule of thumb said that the shorter the drying time, the poorer the varnish, but with the development of synthetic resins, the rule has become subject to many exceptions. To avoid inferior products, buy from reliable dealers.

Penetrating floor sealer is a type of varnish that penetrates the wood and seals it with a thin, hard coating. It is not as tough as ordinary floor varnish, but hides scratches better than floor varnish.

2-14. Stains

A wood *stain* is used to darken or color a wood without hiding the natural grain and texture. It is important to note that a stain is not a protective finish; a clear finish such as one of those discussed in the preceding paragraphs is needed to protect the surface. Stains are made of colored dye dissolved in a solvent and are usually classified by the solvent.

Since stains are absorbed by the wood, they must be applied to bare wood only. The stain is usually brushed on freely, but it may also be wiped on with a cloth. The excess is wiped off with a cloth after the stain has been allowed to soak into the wood for five to fifteen minutes. The stain should be allowed to dry for twenty-four hours before further treatment.

Oil stains use turpentine or naphtha as the solvent in an oil vehicle. These stains are used mainly on exteriors. The oil penetrates so that the wood is colored not only on the surface, but also to an appreciable depth. For this reason, the stain doesn't have to be renewed as often as paint, even if the wood wears.

Oil stains may be clear or pigmented. The pigmented types have a small amount of pigment added which partially obscures the grain of the wood. However, like the clear stains, the pigmented type also soaks into the wood and in this respect is different from paint, which simply covers the surface.

Water stains use water as a vehicle and penetrate much more deeply than other stains. They are difficult to apply because they start to penetrate immediately and may do so unevenly, causing the surface to look streaked. Water causes the grain to rise so that more sanding is needed than with other stains.

Alcohol stains use alcohol or acetone as a vehicle. They are the fastest drying and consequently the most difficult stains to work with. They are used mainly under lacquers. The home handyman should avoid using either water or alcohol stains since considerable practice is required to master the proper application of these materials.

Non-grain-raising stains use a special solvent that does not raise the grain of the wood as it is applied. These stains do not penetrate as far as the others but they do not bleed through protective coatings or fade. Because the grain is not raised, sanding is not necessary after this type of stain is used. This stain is usually applied by spraying, but it can be wiped or brushed on as well.

Varnish stain is a combination of a stain and a varnish. It accomplishes two operations in one step, staining and varnishing the wood at the same time. It is cheap and fast, but is neither attractive nor long-lasting.

After wood has been stained, a *filler* is applied to fill the pores of wood. Fillers come in two forms, liquid and paste. Liquid fillers do not fill as well as the pastes and thus are used only on close-grained woods. Paste fillers are usually thinned with turpentine and may be used on any type of wood. Close-grained woods include pine, fir, maple and birch; open-grained include oak, walnut, ash and mahogany.

The filler is usually brushed on after the wood has been stained. As it dries it becomes duller, and when this happens (about fifteen minutes after application), the excess is wiped off with a clean rag. The surface is then ready for finishing. On close-grained woods having little porosity, shellac can also be used as a filler.

2-15. Enamel

Enamel is a special kind of pigmented paint that uses varnish in the vehicle and dries to a smooth coating that combines the durability of varnish with the beauty of a pigmented finish. In popular jargon the term is applied to any gloss paint, but the gloss paints do not have the wearing qualities of enamels. Enamel is available in flat finishes as well as in a variety of glossy finishes. The amount of gloss in the finish is controlled by the ratio of pigment to oil in the mixture. The more oil there is, the shinier will be the surface finish.

Just as there are many types of varnish, so too there are many types of enamel, depending on the varnish and the vehicle used. Alkyd enamel and latex enamel are the two broad classes that depend on the vehicle. Latex enamel is not as durable as the alkyd variety, but does have the advantage of all latex paints: painting tools may be cleaned in water. Enamels are made for special purposes and are designated as *floor enamel, implement enamel* (for farm implements), *exterior enamel, automobile enamel.* For best results an undercoater should be used with enamel. This is usually a white finish and may be flat; if a high-gloss final finish is required, however, the undercoat should have a gloss finish. The most important requirement of the undercoater is that it form a tight surface film, so that the final enamel finish will not penetrate it. Ordinary paint could be used as an undercoater, but to ensure a tight film it is best to use the enamel undercoater specified for the particular final enamel finish being used. This is usually indicated on the label of the can of enamel. Enamel is generally applied with a brush and dries rapidly.

Painting Tools and Equipment

3-1. Brush Filaments

Paintbrushes are among the oldest of man's tools: primitive man made them of reeds or stems of large leaves, shreaded into thin fibers. Later, animal hair proved superior for brush filaments, and the best was *bristle,* the hair of the hog. The word "bristle" is frequently, but erroneously, applied to all bristle-like material, and it is common to find references to "nylon bristles" or "horsehair bristles". However, the Federal Trade Commission of the United States Government insists that brushes labelled "pure bristle" be 100 per cent hog hair and that brushes with mixed filaments have the ingredients listed in descending order of percentages. Thus, a brush stamped "bristle and horsehair" should have more bristle than horsehair.

The *filaments* or *hairs* of modern paintbrushes are classified as either *natural* or *synthetic.* Natural filaments come from animals; vegetable fibers are no longer used. Although the best natural filaments are bristles, horsehair is also used since it is cheaper. It is markedly inferior to bristle, however. To combine the economy of horsehair with the performance of bristle, some manufacturers make brushes with mixed filaments, bristles on the outside and horsehair in the center. Brushes with mixed natural filaments are

superior to plain horsehair brushes, but are not as good as pure bristle.

Synthetic or man-made filaments are continually being improved. The early synthetic brushes used nylon filaments and were inferior to bristle brushes. However, improved nylon and other newer synthetics are as good as bristle for most applications and far better than bristle when used with latex paints. Natural bristles absorb water and thus lose their stiffness and elasticity in water-based paints. Nylon filaments absorb a small amount of water but not enough to affect their painting performance; newer synthetics like *polyester* absorb almost no water.

Paintbrushes must be capable of picking up a large amount of paint and carrying it with a minimum of dripping. The paint-holding capacity of a brush is determined primarily by flags or split ends at the tips of the filaments. Hog bristles, which have these flags naturally, are excellent paint carriers. However, as a paintbrush is used, the flags wear down, leaving long, *tapered points* on the tips of the bristles. Even so, bristles still hold paint by capillary action between the tapered points and because of their rough surfaces. This roughness extends the full length of the bristle and allows a brush to be used after considerable wear. Horsehair filaments are not split naturally, but manufacturers of brushes sometimes split the ends of horsehairs to make flags and thereby improve the brush. How-

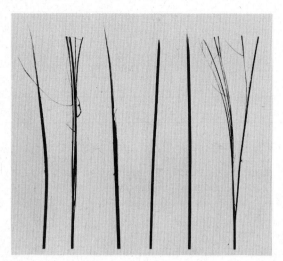

Fig. 3-1. Nylon filaments.

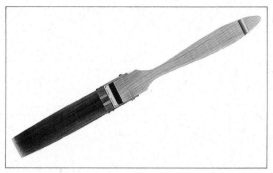

Fig. 3-2. Brush construction.

ever, horsehair lacks the roughness necessary after the flags wear off.

The first nylon brushes used smooth filaments with split ends, but these were not as satisfactory as bristle brushes. Then the nylon filaments were roughened in manufacture and as a result modern nylon brushes give excellent performance. Manufacturers mix long and short as well as flagged and pointed filaments to get improved painting action. The flags and points of some nylon filaments are shown magnified in Figure 3-1.

Filaments must be resilient; that is, when bent and released, they should resume their shape. This elasticity is achieved by tapering the filament from its base to its tip. Hog bristles have this taper naturally, but horsehair does not, another reason why horsehair filaments are inferior. Synthetic filaments are manufactured with tapers and have superior resilience.

Most paintbrushes have black filaments, only because consumers seem to prefer black to any other color. However, natural bristles come in many colors, and synthetics can be made in all colors. To make filaments black, they must be colored with an excellent grade of black dye, which increases the cost of a brush. Cheap dyes could run and ruin a paint job.

Although bristles have natural characteris-

tics that make them especially suitable for paintbrushes, manufacturers have developed synthetic filaments with flags, rough surfaces, tapers and points so that the synthetic brushes are at least equal to the best natural brushes. Moreover, bristles do vary in quality, whereas the quality control in the manufacture of synthetic filaments is excellent.

Since natural filaments, such as hog bristles or horsehair, tend to wilt in water, they should not be used with latex or other water-based paints. Nylon brushes were a necessary invention when latex paints were developed. The early nylon brushes, as well as the cheaper ones still available today, work well with latex, but deteriorate in oil-based paints. However, good nylon filaments and other synthetics manufactured today can be used with oil-base, alkyds and latex without deterioration.

The construction of a brush is illustrated in the cross sectional view of Figure 3-2. The filaments are attached to a wooden or plastic handle by means of a metal ferrule that holds them firmly in place. The part of the filaments next to the handle is called the *heel* of the brush. Shown in the photograph are spacer plugs in the heel to give the brush a taper and to allow it to hold more paint. In their inferior models some manufacturers achieve a taper by omitting filaments in the center of the clump, so that the brush is "hollow". Such a brush is difficult to use since it does not release paint evenly.

The most important consideration in the manufacture of a brush is the blend of filaments of different lengths and sometimes different types to achieve the best painting action. Some manufacturers insist that their formulations are best, and keep them closely guarded secrets. However, many manufacturers make excellent brushes, and although the proper blend of long and short filaments is important, it is not a mysterious process. In good brushes, filaments are selected to give the right amount of stiffness and flexibility while preserving maximum paint-carrying ability.

3-2. How to Select a Brush

Don't buy a cheap brush. A good brush is an expensive but good investment. A cheap brush is difficult to use and does not give as good a paint job. A good brush carries more paint and does not drip as a cheap brush does. With a good brush you will paint more rapidly since the paint is applied evenly without brush marks. Cheap brushes wear out rapidly or become unmanageable because of "wild" bristles, but a good brush will last many years; in fact, some manufacturers offer nylon brushes with a lifetime guarantee if they are "cleaned and maintained according to our instructions".

Excellent brushes are available with either natural or synthetic filaments. You should buy synthetic brushes both because they are superior for use with latex paints and because they are easier to clean when the painting is finished. The most common synthetic material in brushes is nylon, which is quite satisfactory. Polyester is somewhat better, and undoubtedly newer synthetics will be developed which will improve performances. Meanwhile, any good synthetic brush will suit the average home owner who paints only occasionally.

Make sure the brush is full; that is, not hollow in the center. The filaments should be held tightly by a ferrule that is fastened securely to the handle. The ferrule itself should be made of stainless steel or some other metal that will not rust when used with the new water-base paints.

The handle should be made either of wood or plastic, preferably the latter since it is not affected by water. Even if you plan to use the brush for oil-base paints, you should choose a plastic handle because you will wash the brush in water when you are finished. The only criterion you need apply to the shape and the size of the handle is whether or not it feels comfortable when you hold it.

Be sure the filaments of any paintbrush you buy are tapered and have flags. The taper makes the brush flexible and elastic. Press the brush against the palm of your hand or on a smooth surface and observe the filaments. First, they should not fan out at random, but should remain in a more or less compact cluster. This is an important consideration when you are painting in corners or on trim, where wildly spreading filaments can apply paint to the wrong areas. Second, the filaments should flex more at the tip, where they are narrower, than at the base. If filaments bend at the base, but are more or less straight, they lack the necessary flexibility to apply paint smoothly and evenly. Now release the filaments. They should spring back to their former position. Beware of brushes with filaments so soft that they feel silky and do not return when released or so hard that they feel hard and sharp against the palm of your hand.

The proper length of filament depends on the size of the brush. A single filament is flexible, but when many filaments are bound together, their flexibility varies with their lengths and number. Typically, a good 4″ brush has filaments about 4″ long, as shown in Figure 3-3. Wide brushes, like the 6″ brush in Figure 3-4, also have filaments about 4″ long. Narrow brushes, like the 2″ brush in Figure 3-5, have filaments longer than the width of the brush.

If you have any doubts about which brush to choose, ask your dealer to recommend one. Any high-quality synthetic brush made by a reputable manufacturer should be quite acceptable. Let your paint dealer be your guide if you don't know who the leading manufacturers of paintbrushes are.

For painting large surfaces, you may use a *wall brush* or a roller. Wall brushes range from 3″ to 6″ in width. If you decide to use a brush

Fig. 3-3. Four-inch brush.

Fig. 3-4. Six-inch brush.

rather than a roller, you should buy the largest that you can handle comfortably. The bigger the brush, the faster the job will be completed since you cover more area with each stroke. However, this is true only if your arm doesn't tire from wielding a large brush heavy with paint. The 6″ model shown is too big for the occasional painter, and most home owners find a 4″ brush, like that shown, a convenient size to use.

Regardless of whether you use a brush or a roller for large areas, you will need in addition a small brush for painting trim, woodwork and small areas such as the corners at the junction of walls and ceiling. These small brushes are called *trim* or *sash* brushes. The name "sash brushes" would seem to indicate that this kind of brush is smaller than a trim brush, but since one brush can be used for both sashes and trim, some manufacturers lump them under the name sash and trim brushes. Sash brushes run from about 1/2″ to 1-1/2″ in width, and trim brushes from 1″ to 2″. Trim brushes are small versions of the larger wall brushes, but sash brushes may have round or oval cross sections as well as rectangular (see Figure 3-6). The flat trim brush in (a) is like the 2″ brush shown in the preceding illustration. The cross section is rectangular and the filaments are cut square. The oval sash brush in (b) usually comes with a round handle, so that you can hold it as you

Fig. 3-5. Two-inch brush.

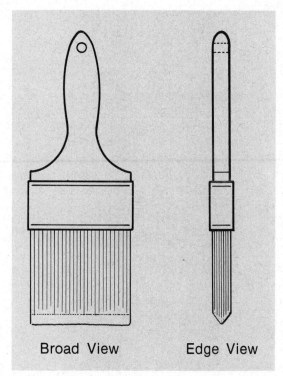

Broad View   Edge View

Fig. 3-7. Varnish brush.

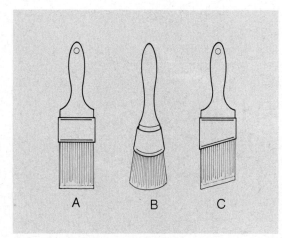

Fig. 3-6. Sash and trim brushes.

do a pencil. The angled sash brush in (c) has a rectangular cross section, but the filaments are cut at an angle to enable the user to paint window sashes without bumping the glass.

A varnish brush (Figure 3-7) is a special brush for applying varnish or enamel. It has a chisel edge, as shown in the edge view. The filaments are longer and softer than those of a trim brush. The chisel edge enables you to apply paint or varnish in corners or on furniture or trim without accidentally touching surfaces that are not to be painted.

If you were a professional painter, you would want a wide variety of brushes for different purposes, but for the occasional paint job, you can get by with a wall brush or a roller for larger areas and one small brush. It's not wrong to use a varnish brush for painting trim or a trim brush for varnish. Whether you use an angled sash brush or a flat brush or an oval brush is a matter of personal preference.

A trim brush without bristles (Figure 3-8) is also available. It is made of plastic sponge and looks and feels like a paintbrush. Originally, plastic sponge brushes were thrown

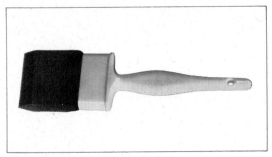

Fig. 3-8. Plastic sponge brush.

away, but now that they are made with improved plastics, they can be cleaned and used again. They apply a smooth finish without brush marks. Note the chisel edge that permits painting of sharp, straight edges. A plastic sponge brush can be used with varnish, enamel, latex, alkyd or oil-base paint. It should not be used with lacquer or shellac.

3-3. How to Use a Brush

During painting and cleaning and especially during interruptions, it is necessary to hang a brush in a container of solvent with the tip off the bottom. The simplest way to do this is to pass a wire, such as may be cut from a metal coat hanger, through a hole in the brush. With the wire in place the tip of the brush should be about an inch from the bottom of the can. A suitable container for solvent is a large coffee can or a fruit juice can or an inexpensive paint bucket. Most brushes have holes in their handles as is evident in Figures 3-3, 3-4 and 3-5, but these holes are not always located in the right place for the container you have on hand. The first thing to do with a new brush, then, is to drill a hole in the handle at a suitable location for supporting it in the can.

Before you paint with a new brush, you should clean it and prepare it for the job. Tap the brush vigorously against the palm of your hand and snap the filaments against your fingers to shake out loose filaments and dust. If the brush has natural bristles it should be suspended in a can of linseed oil for at least 12 hours. This softens the bristles so that the

brush holds more paint. After removing the brush from the linseed oil, squeeze out the excess oil, and swish the brush in turpentine. Note that the linseed oil bath is a one-time event: after the bristles have absorbed the oil, they will not need conditioning again. If you are going to use the brush with alkyd paint, you can omit the turpentine wash. For use with latex paint, wash out the turpentine with soap and warm water, then rinse. A word of caution: since linseed oil is not compatible with shellac or lacquer, skip the oil pre-conditioning if you are going to use your brush with these materials.

If the new brush is a synthetic like nylon or polyester, it doesn't need preconditioning. You may want to wash the filaments in warm water and mild soap to make sure that all traces of dirt and oil are removed. The brush is now ready for painting. If you are using latex paint, you don't have to wait for the brush to dry out.

Occasionally a new brush will have one or two filaments that are "wild"; that is, that do not lie parallel with the rest. Simply cut them off with a sharp knife.

Hold the brush in any way that feels comfortable. You may find that small brushes are more easily controlled by holding them with the fingers, like a pencil. For large surfaces, where delicate control is not needed, it may be less tiring to hold the brush in your fist.

Do not overload the brush. Too much paint in the brush tends to drip and splash when applied and leaves a thick film on the surface. To avoid this, dip just half of the filament in the paint and remove excess paint by tapping the brush lightly against the inside of the container. Do not drag the brush across the rim of the can, since that tends to make the filaments bunch together. Never stir the paint with your brush.

Try to avoid getting paint in the heel of the brush. This is the most difficult part of the brush to clean, and if any paint remains and hardens in the heel, it shortens the flexible part of the bristles and makes painting more difficult.

Paint should be applied with short strokes and then smoothed out with longer ones.

Don't use too much pressure. Lift the brush at the end of each stroke before reversing direction. Never paint with the side of a brush since it causes curling. For smaller areas, use a smaller brush. Don't use a large brush to paint pipes and other objects that are thin compared to the width of the brush. Don't jab the brush into corners. The emphasis should be on letting the brush do the work. Let the paint flow onto the surface instead of forcing it.

For smoother finishes, go over the surface lightly with the brush unloaded after every three or four brushfuls. The smoothing strokes with the brush almost dry should be at right angles to the direction of the first strokes.

If a loose filament from the brush is deposited on the freshly painted surface, you can usually pick it up with the brush by dabbing at it lightly. After removing the filament, brush lightly over the area to cover the marks.

Interruptions pose a problem. If you stop for a short time, you may suspend the brush in the paint bucket, making sure that the filaments are not immersed more than half their length and that they do not touch the bottom of the container. The brush should not be left in paint for more than an hour. For very short interruptions (answering the telephone), lay the brush flat on a pile of newspapers. However, if you are using latex paint, you must realize that it dries quickly, and you must not leave the brush exposed to the air for more than a minute or two. You can wrap the brush in a cloth dampened with water and lay it flat, and it will not dry out as long as the cloth remains moist. For longer interruptions, you can wrap the brush in a double thickness of aluminum foil or a transparent plastic wrap to keep it from drying out.

If your paint job takes more than a day, and it is necessary to store the brushes overnight, your procedures will depend on the type of paint you are using. With latex paint, cleaning is simple: you should clean out the paint from the brush with soap and water. Lay the clean brush on a flat surface. It is not necessary to wrap it or dry it. With alkyds or oil-base paints, you can suspend the brush in linseed oil or turpentine overnight. The thinner can cover the filaments completely. When you are ready

to paint again, wipe out the excess thinner and rub the brush on newspaper until it seems dry.

3-4. How to Care For a Brush

The most important part of maintaining a paintbrush is proper cleaning. Though not a difficult task, cleaning is tedious and while you may be tempted to rush or do the job incompletely, you should make sure every bit of paint is removed from the brush before you store it.

To clean a brush, soak it in thinner or solvent for a few minutes. Make sure the thinner works up into the heel of the brush. Then wash the brush with water and mild soap. That's all there is to it, but you must be thorough. To ensure that the thinner gets into all parts of the brush, especially the heel, work it in with your fingers. You may want to wear rubber gloves if you are using turpentine or other solvents that can dissolve the oils in your skin. Agitate the brush in the thinner so that every filament is separately exposed to the solution. The procedure varies slightly with the kind of paint involved.

After painting with latex or other water-base paints, rinse out as much of the paint as you can under a water faucet. Work the filaments with your fingers to get out as much as possible, especially around the heel of the brush. Mix a solution of mild soap and warm water and soak the brush in it, again working the filaments to get the soapy water well into the heel. Rinse several times in clean warm water or under a warm water faucet until all paint is removed.

After using alkyds or other oil-base paints, soak the brush in kerosene and agitate it to get the solvent into all parts of the brush. Repeat until the brush shows no trace of paint. Wash in soap and warm water until there is no sign of color in the wash. Then rinse in clear water.

For other finishes, use the thinner or solvent

specified by the manufacturer. Always follow with a soapy water bath to wash out the thinners and a clear rinse to remove the soap.

When the brush is clean and still damp from the water rinse, comb out the filaments with a brush comb. The brush comb consists of a wooden or metal handle with metal pins or teeth. Combing straightens interior filaments that might be matted and keeps the brush in proper shape.

Hang the brush up to dry with the filaments pointing down. When it is dry, wrap it in foil or plastic to keep out dirt, and store it flat.

Though hot or warm water is fine for use on all synthetic brushes, do not use it on natural bristles; use cold instead. Some synthetics cannot be soaked in alcohol or lacquer thinner. Find out when you buy your brush whether it can be used in shellac, lacquer and their solvents. Ammonia can be used in place of alcohol as a solvent for shellac.

Outside of proper cleaning, the maintenance of paintbrushes lies chiefly in avoiding improper use. Some of the common offenses are mentioned in the preceding pages, and repeated bad practice can lead to the troubles illustrated in Figure 3-11.

Wild filaments, illustrated in Figure 3-11 (a) are the result of improper cleaning and failure to comb out the brush when done. The fishtail in Figure 3-11 (b) is caused by painting a pipe or other thin structure with a brush too wide for the job. Fingers shown in Figure 3-11 (c) are usually the result of using the brush edge-on instead of painting with the broad side. Fingering can also be caused by drawing the brush frequently across the rim of the paint container. Curl in Figure 3-11 (d) is caused by letting the brush rest with its filaments in contact with the bottom of the container. Other troubles are the result of improper cleaning, leaving paint in the heel of the brush or neglecting to clean a brush at all, so that it dries to a solid, hard lump.

If you have brushes that have been damaged by neglect or misuse, don't throw them away. It is usually possible to get a good brush back into shape, assuming its filaments are still firmly attached to the handle. The treatment consists of two steps: (1) removing all traces of paint, and (2) shaping the brush. Dried paint in a brush must be removed before it can be reshaped. The troubles depicted in the figures can be accompanied

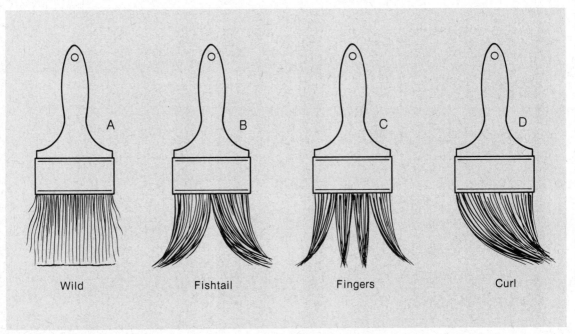

Fig. 3-11. Neglected brushes.

Before applying spackle to a plaster wall, soak the wall around the crack with water.

by dried paint in the brush or they may occur even when all paint has been removed by careful cleaning.

If the brush is a solid mass of dried paint or if it has a lot of paint dried in the heel, it should be soaked in a chemical paint cleaner, which is stronger than most paint thinners. Hang the brush in a container so that the tip does not touch the bottom and pour in enough cleaner to cover the filaments and the ferrule completely. For a heeled-up brush, a few hours soaking may be enough, but if the brush is completely solid with paint, you may have to let it soak for a week. It is better to buy a mild cleaner that takes longer since it is less likely to damage if you soak the brush too long. After soaking, the paint should have softened to a jelly or completely dissolved. Comb out any loose paint with an old comb and then wash the brush in hot, soapy water. (Bristle brushes cannot stand hot water, so must be washed in cold water and strong soap or detergent.) Repeat, if necessary, until the bristles are soft and flexible, as in a new brush.

After removing all traces of old paint from the brush, it may still be misshapen as shown in Figure 3-11. Suspend the brush in linseed oil for about an hour and then comb it as straight as possible with the brush comb. Combing straightens interior filaments that might be matted. Dip the brush in linseed oil again and while it is still wet, shape it and wrap it tightly in aluminum foil to maintain the shape. Store it flat for at least two days. Then clean out the oil with turpentine, and wash the brush in warm, soapy water to wash out the turpentine. Rinse in clear water and comb the brush once more. Hang to dry and then wrap in plastic wrap or foil for storage.

3-5. Types of Rollers

For painting large, flat surfaces, paint rollers are easier and faster to work with than brushes. Like brushes, rollers come in a wide variety of styles, sizes and materials. Also like

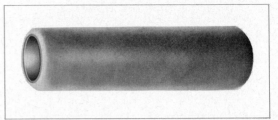

Fig. 3-12. Roller cover.

brushes, prices of rollers vary depending on material and quality of workmanship.

The cover of a roller is shown in Figure 3-12. Covers may be made of synthetic material like nylon or rayon or of a natural fiber such as wool or mohair. Natural fibers should not be used with latex or other water-base paints, but mohair provides the smoothest finish for applying enamel. Rayon rollers are cheap and generally not worth the money. Good quality nylon rollers can be used with any kind of paint.

The core of the cover, the inner tube, may be made of laminated cardboard, metal or plastic. The only important consideration is that the core retain its shape after washing in thinner and water. Cardboard tubes are cheap, but do not last as long as metal or plastic cores. If the core is metal, it should be rustproof, since it will be used in water-base paints and will be washed in water.

The nap on the roller may vary in length from about 1/4″ to well over 1″. The shorter the nap, the smoother the coating you can apply. For most smooth paint jobs, use a roller with a 1/4″ nap. For slightly rougher surfaces like plaster or gypsum board, a 3/8″ nap is better. You rarely need anything longer than this for indoor painting. The 3/4″ nap is usually used for outdoor finishes on rough surfaces like stucco, rough plaster and rough wood. Longer naps may be used on rough masonry and chain-link fences.

Rollers vary in length from about 3″ to 18″. The larger the roller, the greater the surface you can paint with each stroke; large rollers may be awkward to handle, however, and require oversized pans to hold the paint. For most indoor painting, a 9″ roller is adequate.

The roller cover slips onto a frame such as

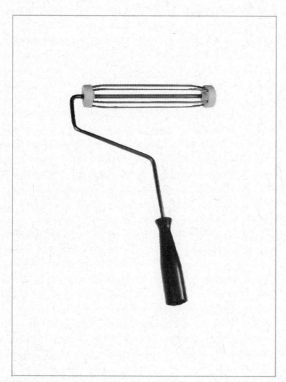

Fig. 3-13. Roller frame.

spot. To reach these inaccessible areas, special corner rollers have been designed. They are usually made of polyurethane or sponge rubber foam.

3-6. How to Select a Roller

It is possible to paint using only brushes, as professionals have been doing for years, but a roller does simplify painting large areas. If you decide to buy a roller, don't look for bargains. A cheap roller is a poor investment unless you are planning only one paint job and intend to throw the roller away when you are through.

   Choose a good synthetic cover since you will probably use the roller with latex paint. Avoid cheap rayon covers. The nap should be 1/4″ high for most indoor painting, but if you

that shown in Figure 3-13. An assembled roller is shown in Figure 3-14. The roller part of the frame should rotate easily, but not so fast that it will spatter paint. You should be able to attach the cover to the frame quickly and easily so that you can simply change covers when you want to use a different paint. The handle should be made of plastic or some other material that can be washed in water. For painting ceilings without a ladder, a roller with an extension handle is desirable. The extension screws into a threaded socket in the handle of the frame.

   Rollers are used with a paint pan that has a sloping bottom. Pans vary in width to accommodate rollers of different sizes; the most common, shown in Figure 3-15, is designed for a 9″ roller. Legs at the shallow end of the pan keep it from tipping and also keep the upper end of the ramp out of the paint. Pans are also available with clamps to fasten on an extension ladder for outdoor painting.

   When a roller is used at the junctions of walls and ceiling it cannot always reach every

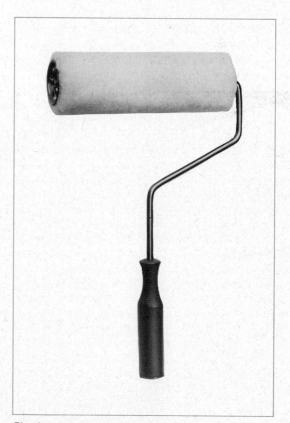

Fig. 3-14. Assembled roller.

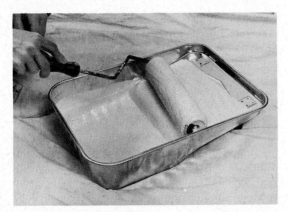

Fig. 3-15. Roller and pan.

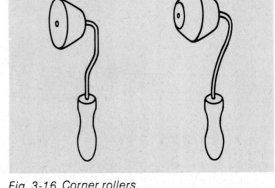

Fig. 3-16. Corner rollers.

are going to paint over plaster, gypsum board or wallpaper, get a cover with a 3/8″ nap also. Make sure both covers have a core that will not deteriorate when washed in warm water. The covers should be 9″ wide.

The frame should be made of aluminum or stainless steel and should have a plastic handle. Again, the main consideration is the frame's ability to withstand corrosion in water.

You should be able to slide either cover on the frame quickly and firmly, and once attached, the cover should rotate easily. If you don't like standing on a chair or ladder to reach the ceiling and upper parts of the walls, you should get a frame with a detachable extension handle.

The kind of pan you buy is not a critical matter. Pans are usually made of aluminum, which is lightweight, inexpensive and rustproof. For indoor painting, you will not need clamps to attach the pan to a ladder.

Small corner rollers (Figure 3-16) are convenient if you don't have any brushes, but more difficult to use and clean than a brush. Don't buy one if you have a trim brush.

3-7. How to Use a Roller

A new roller should be conditioned before use. Wash the cover in soap and water to remove all dirt. After rinsing, you can use the roller immediately with latex paint, since it can be wet when applying water-base paints. For alkyd paints, let the cover dry thoroughly before using the roller.

Paints that dry quickly, such as lacquers and fast-drying enamels, cannot be applied with a roller, but most other paints can be rolled on. If you have any doubt, read the label on the can. In addition to telling you whether the paint can be rolled, the label will also supply special instructions for rolling. Many paints can be rolled as they come from the can, but some need to be thinned. The type and amount of the thinner needed are always specified on the label.

After the paint is well stirred, pour enough in the roller pan to cover about half of the sloping part of the bottom. Every time you have to refill the pan you should stir the paint before pouring it. Roll the paint roller in the paint in the deep part of the pan until the roller is filled evenly. Roll the roller on the exposed slope, as shown in Figure 3-15, to remove excess paint. Apply the paint to the wall, ceiling or other flat surface first in three or four strokes in a random pattern. This averts the undesirable tendency of the brush to deposit all its paint in one spot. Alternatively, make a large W with the first four strokes. Now roll the roller across the first strokes, spreading the paint uniformly on the surface. Repeat, making your next random pattern or W a short distance away from the portion already finished. Blend in with horizontal strokes.

To avoid stooping and reaching, place the pan about 2′ or 3′ above the floor. A small

Edging glider used on edge of a door frame.

table or box, covered with protective cloth or newspapers, can be used. To reach a high ceiling, use an extension handle.

Do not press too hard, as that causes the roller to paint unevenly. Roll it slowly and gently until the roller begins to get dry. Then dip it again and continue.

Cutting in is painting in the corners that cannot be reached by a roller. This can be done with a special corner roller, but it is simpler to use a brush, dipping it in the roller pan. Ideally, one person uses a brush for cutting in (as shown in Figure 3-19), while a second uses a roller on the flat surfaces. Cutting in should be done before rolling as the roller can blend the paint onto the cut-in section without brush marks. In the figure shown the ceiling is painted the same color as the walls, but if a different color is used, some form of masking or careful trim painting is required. This is discussed in the section on painting specific parts of a room.

If you are interrupted while using a roller, simply put the roller into a container of solvent (water, if latex paint is being used). When ready to resume painting, roll the roller back and forth on some old newspapers until the solvent is almost dry, and then continue painting as before.

Fig. 3-19. Teamwork painting.

warm soapy water and then rinse out the soap. Hang up to dry. Wrap the cover to keep off dust until the next job.

3-8. How to Clean a Roller

If you clean a roller as soon as you finish painting, it is a very simple task. The longer you delay, the more difficult it gets. Begin by rolling it on old newspapers to remove as much paint as possible. Pour some solvent into the tray used for the paint. For latex paint, use warm water; for alkyds, use turpentine or mineral spirits. Rinse the roller cover in the solvent. When the paint is almost all gone, remove the roller cover from the frame and wash each separately in the solvent. For latex paint you may find it simpler just to rinse the tools under warm water. For a final cleaning, no matter what solvent was used, rinse in

3-9. Other Applicators

Since a napped surface makes a good paint applicator, paint manufacturers supply other convenient accessories for special painting requirements. These special applicators are available with synthetic or natural naps and in a variety of shapes.

A handy tool for painting stairs is a *paint glider,* shown in Figure 3-20. The pad is flat and is removable from the handle. The handle may be adjusted to any angle and locked there for ease in painting. The glider shown has a mohair pad and is about 6″ by 3-3/4″; larger pads are also available. For tight places and stairs, where a roller is inconvenient, the glider is a useful addition to your set of painting tools. The nap side of a nylon fiber glider is shown in Figure 3-21.

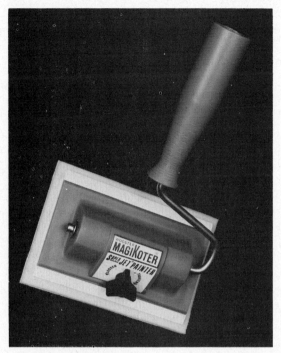

Fig. 3-20. Paint glider.

Fig. 3-21. Nap side of glider.

A glider may be used to paint a large wall and is faster than a brush, but slower than a roller. In Figure 3-22, paint is being applied to a wall with a glider, starting with the same W pattern that is used with a roller. In general, however, you should use a roller for walls and reserve the glider for stairs, woodwork and cramped quarters.

A special glider for sharp edges is shown in Figure 3-23. A small wheel or projection at the side is held in contact with a window frame, corner, or other surface that is not to be painted, and this keeps the pad itself away. Paint is deposited in a straight line with no danger of getting any paint on the clean surfaces. In the illustration, the glider is being used to paint the inside stop of a double-hung window without getting paint on the sashes.

Another tool with a nap is a *painter's mitt,* a large mitten covered on the outside with a nap of wool or nylon. When using it, your hand becomes the painting tool. There is nothing better for painting pipes, posts, gutters, and many other round or irregularly-shaped objects.

A roller pan is the best paint container to

Fig. 3-22. Glider on wall.

use with a glider or mitt. When using these applicators, follow the same thinning instructions as those given for use with a roller. Dip the nap into the paint, wipe off the excess paint on the sloping bottom of the pan, and apply the paint to the surface much as you would with a roller. After the job is done, the glider or mitt should be cleaned immediately in the same manner as a roller is cleaned (as described in Section 3-8.) For most indoor

Do not press a roller too hard, as that causes it to paint unevenly.

When painting with a spray can, keep the nozzle at a constant distance from the surface.

Fig. 3-23. Edging glider.

home painting you need only a brush and a roller. If you buy a mitt for painting pipes or a glider for stairs, it should have a nylon nap since this is best for use with latex paint.

3-10. Spraying

Another way to apply paint is by spraying. Paint may be purchased in a spray can for quick touch-ups or for a small paint job. There is no cleaning up afterward. For large jobs, a spray gun is used. It is not necessary to buy a gun, since most paint shops will let you borrow or rent one when you buy the paint. Ask your paint dealer to set the controls for the paint and the job.

Paints must be thinner for spraying than they are when applied by brush or roller; they should be thinned with appropriate solvents. Once the mixture is in the gun, spraying is very simple. You can spray almost any kind of paint on almost any surface, including cloth,

leather, paper, metal, plastic, masonry and wood. For wicker furniture and odd-shaped pieces, spraying is the fastest way to paint, but extra precautions are necessary. It is almost impossible to control the *overspray,* that is, the spraying that goes beyond the object being painted. Thus, surrounding objects must be protected more thoroughly than they are when a brush or roller is being used. Adequate ventilation is necessary, and even if you are painting outdoors you should wear a mask. For long jobs, you should also wear a cap and gloves.

Mix and strain the thinned paint before putting it in the gun. The spray gun or can must be moved so that (1) the spray is perpendicular to the surface being painted and (2) the nozzle remains at a constant distance from the surface. If the surface to be painted is curved, as, for example, the outside of a barrel, then the spray gun must follow a curved line to remain equidistant from the surface. The tip of the nozzle should be about 8″ from the surface. (This may vary with different spray guns and different aerosol cans, so read the label before using the equipment.) Before you begin painting the surface, you should practice on a piece of cardboard or wood scrap. As you move the gun over the surface, it should spray continuously. Do not start or stop the spray in the middle of a stroke. Thus, you start the spray with the gun pointing slightly off the surface. Then move the spray all the way across until the spray is off the surface at the other side. When coming back, always overlap your strokes. There will be no lap marks.

Always save the best for the last: when spraying furniture, start with the undersides or other inconspicuous places. Finish on the best surface. The reason for this is that it is impossible to avoid some misting. If you did the top, the misting from the rest of the job might make the top look sandy. Rubber-based paints cause less misting than others.

To clean the spray gun, empty the paint tank and wash it in the proper solvent for the paint you used. Then fill the tank with solvent and spray it through the gun until it comes out clear. Return the clean spray gun to your dealer.

3-11. Other Accessories

Before you begin painting, you should make sure you have any accessories you may need. You may not need all of these items for every paint job, but it is better to have them on hand than to have to interrupt your painting to look for one.

- Brushes

- Roller and tray and other applicators

- Paint buckets and stirrers

- Stepladder

- Drop cloths

- Newspaper and clean rags

- Putty or other fillers

- Paint shield

- Masking tape

- Thinners

- Sandpaper

- Putty knife

- Screwdriver

- Hammer and nailset

- Rubber gloves

- Painter's hat

Some of these items are needed in preparing the surface, others in actual painting, and some for cleaning up afterward.

You will certainly need a stepladder or something to stand on to reach the junction of walls and ceilings. Even if you have an extension handle on your roller, you will still have to cut in with a brush. You can stand on a solid chair or table, but do not risk standing on a box on top of a chair. A convenient platform can be made from a plank and two chairs which will enable you to paint a large part of the ceiling with a minimum of climbing up and down. Make sure the board is solid enough and wide enough to hold you.

No matter how careful you are, you will drip paint and hence will need some sort of protective covering for furniture and floors. Drop cloths made of canvas, cloth or plastic are available. In a pinch you can use old newspapers. When you have covered the areas to be protected, you don't have to worry about dripping and thus can paint more freely and with greater speed.

For painting near a surface that is not to be painted, you need masking tape or a paint shield. Inexpensive metal or plastic shields are available, but you can get by with a piece of ordinary cardboard.

To complete a painting job satisfactorily, you need some common tools: (1) a hammer and nailset for driving nails below the surface; (2) a putty knife to spread putty or filler in cracks and over nailheads so that the surface is smooth and unbroken; and (3) a screwdriver for removing hardware that is not to be painted. You will need containers for mixing paint and for holding solvents when you clean your painting tools. You can use old coffee cans or paint cans; the latter are especially handy since they have handles. If you want to save old paint cans for this purpose, clean them with the appropriate solvent as soon as you have used up the paint. Cans that held latex paint can be washed in water. If you want to avoid a large part of the cleaning process, buy inexpensive cardboard paint buckets and throw them away when the job is finished.

Even if you have drop cloths, you should have some old newspapers to wipe rollers and brushes. A good supply of torn sheets and other clean rags enables you to wipe up spills and misplaced paint easily before the paint hardens. Rags are also useful for wiping your hands and face. To protect your hair, it's best to wear a painter's cap; protect your hands with plastic or rubber gloves if you handle irritating chemicals.

Getting Ready to Paint

Paint is not permanent. Eventually a painted surface shows flaws or appears dingy. After you decide (1) that your walls really do need a new paint job and not just another washing, and (2) that you need to paint either the walls alone or the walls and the ceiling both, then the most important aspect of the job will be proper preparation of the surface. A properly prepared surface will hold its paint without flaws three or four times as long as paint applied with little or careless preparation.

In addition to preparing the surface, however, "getting ready to paint" also implies having all tools and accessories at hand so that work can proceed without interruptions. It also means arranging protective coverings, moving furniture, and anticipating all the small chores that must be done when a person is painting.

over it: if the surface feels too oily or waxy, wash off the grease with warm water and trisodium phosphate (TSP) or any household detergent. Wash off the detergent with clear water and wipe dry with a sponge.

If the wall is not greasy, but does have dirt marks or stains on it, it is usually not necessary to wash it since the paint will cover the dirt. One exception is a greasy stain. The grease must be washed off, although it is not necessary to remove all traces of the stain.

Loose dirt or dust in a room can be a problem when it is stirred up by painting. The dust will settle on the wet paint and spoil the finish. If the walls and other areas to be painted are in good condition, dusting and vacuuming may be sufficient surface preparation. Don't forget to dust the tops of doors and window ledges. Dust in any part of the room can spoil a paint job.

4-1. Grease and Dirt

Grease must be washed off before painting begins because paint will not adhere well to grease. Although washing is especially necessary in kitchens where cooking vapors deposit droplets of grease on everything, oils from finger marks can be present in any room in the house. Before painting any surface, then, check for grease by running a finger

4-2. Nails, Holes and Cracks

Preparing the surface involves sealing both holes and cracks, and covering nails so that their heads do not show through the paint. Before cleaning the surfaces to be painted, check to see if any other work is necessary.

Use warm water to wash latex paint from a roller.

Save the cleaning for the last step before painting.

In extremes of heat and cold, the expansion and contraction of a house are apt to loosen nails so that nailheads become visible. This occurs most often in houses where the walls are made of gypsum board. If the nails are simply driven back in the same holes, they will not hold as firmly as before since the original nail holes are now worn. On walls where nailheads become visible, the nails should be hammered back in, but a few extra nails should be added between those already in place in order to strengthen the bond. The nailheads should be driven into the gypsum board so that the hammer dents, but does not break, the paper covering. The nailhead is then below the surface of the wall. Fill the dent with spackling compound, and sand it smooth when it dries. When you clean, make sure you dust off the loose spackle. Now you can paint over the surface and the nailheads will not be visible.

Spackle or spackling compound is available as a white powder which is mixed with water to form a paste. Vinyl-based patching materials which are already mixed and ready to apply can also be purchased. While the vinyl-based mixture is more expensive, it eliminates the worry about getting the proper ratio of water to powder. Latex paint can be used over either compound. For other paints, a primer is usually required.

When a floor squeaks, it is usually a sign that nails have worked loose. Drive a few flooring nails into the area where the squeaks occur. The nailheads should be driven below the surface with a nailset, and the holes above the heads filled with putty or plastic wood. When the seal dries, you can paint over it, hiding the nailhead.

Cracks in plaster or gypsum board walls should be filled with spackle or ready-mixed sealer. Follow the directions on the can. Before applying spackle to a plaster wall, soak the wall around the crack with water. Dry plaster tends to absorb moisture rapidly, and if it is left dry, it draws the water out of the spackling compound rapidly, causing it to crumble. Soaking is not necessary with ready-mixed sealers. For small cracks simply apply the filler with a putty knife, as shown in Figure 4-1.

For large cracks you must make sure the filler is bonded to the plaster. Clean out the hole, removing all loose plaster, and cut the edges to form an inverted V all around the opening. The V holds the dried filler in place. Larger cracks may have to be filled in two stages. First, fill the crack to about an eighth of an inch below the surface. Then, when this is dry, add enough to complete the job. This two-stage approach allows for shrinkage during drying.

For larger holes, such as might be present if you removed an electric fixture, first cut a plug of wallboard to the shape of the hole. It should be of such a thickness that when pressed against the studs or lathing, its surface is slightly below the surface of the wall. Attach the plug to the stud or lathing with a nail and patch on top of it to fill the hole.

After all cracks have been filled and patches sanded, you can clean the room. Loose dust from sanding and old plaster chips should be removed in the cleaning process.

4-3. How to Remove Old Paint

For most paint jobs you can paint over an old coat, just making sure the surface is clean. You can even paint over high-gloss enamels, which normally provide a poor bond to a new coat of paint, by roughing up the surface slightly. The easiest way to do this is to wash the surface with a saturated solution of trisodium phosphate (TSP). This takes off the gloss and washes the wall at the same time.

You cannot paint over calcimine, white-wash and similar water-base washes, since paint won't stick to these coatings. You can remove any of these coatings with TSP or a detergent and warm water. Use a scrubbing brush dipped in the solution and scrub vigorously until the calcimine or other wash is removed.

If the old paint is cracked, flaking, or

Fig. 4-1. Patching holes in wall.

otherwise damaged, you must remove old paint before applying a new coat. There are many ways to remove old paint: scraping, sanding, sandblasting, application of heat, and chemical treatments. Your choice depends on the surface, the amount of time and effort you wish to put into the job, and cost.

Scraping is done with a tool called a *scraper* which is simply a steel blade clamped to a wooden handle. In operation you pull the blade across the paint, pressing down hard as you do so. The blade gets dull quickly and must be sharpened frequently. You can also scrape with a putty knife. Scraping is inexpensive, but it is a lot of work. You should use this method only on small surfaces that can be reached in no other way.

Sanding is used only on wooden floors, using special power sanders that can be rented for the job. Although you could conceivably remove paint from woodwork by sanding, it would be too tedious a job.

Sandblasting is used to remove paint from large masonry surfaces. It is almost never used indoors, although you could rent equipment to sandblast paint from walls and floor in a basement. It would be better, however, just to paint over the old coat, since rental of sandblasting equipment is expensive, and when used indoors, requires use of a mask.

Heating is an easy, relatively inexpensive method of removing paint from large surfaces. Heat may be applied with a blowtorch, propane torch, electric heater, or even with an infra-red lamp. The paint is not burned off, but rather simply heated until it is soft; it may then be scraped off easily with any convenient tool.

Heating is dangerous, however, since there is a danger of fire. Torches with direct flames can ignite or scorch the surface underneath. Electric heaters may also cause a fire by overheating. Lamps are not too dangerous, but they are not very effective sources of heat.

To use a torch or electric heater, apply the

A freshly painted room.

heat without allowing any actual contact between flame or heater and the surface. When the paint begins to bubble, move the heat source with one hand while you scrape the soft film off with a putty knife held in the other. Some electric heaters have a built-in scraper so that in effect the heater and scraper are held in one hand.

Chemical paint removers are the most popular agents for removing paint. They are expensive and are not generally used for large walls, but then, in most cases it is possible to repaint large walls without bothering about paint removal.

WARNING: Chemical paint removers are poisonous, and may be flammable. When you use them you should wear gloves and protective glasses and make sure the room is adequately ventilated.

Chemical removers come in two forms, paste or liquid. Liquid removers are faster and can reach into crevices, but cannot be used on vertical surfaces. Paste can be used on any surface, but is slower-acting than the liquid. Some removers have a water base. After using the chemicals wash the surface with water. Wax is sometimes added to a remover to retard evaporation. After a remover containing wax is used, all traces of the wax must be washed off with turpentine or a similar solvent, and then the turpentine must be washed off with water if latex paint is to be used. The turpentine can remain if alkyd paint will be used.

Make sure you read the label before using the chemical, and heed all warnings. You can apply the remover with an old brush or a rag. Let the remover soak in and soften the paint until you can wipe it off with a piece of rough cloth or steel wool. On walls you can use a putty knife or spatula, but they must be cleaned frequently by wiping them with old newspapers.

4-4. Preparing Wallpapered Walls For Painting

You can paint with latex paint directly over wallpaper. Alkyd paints, however, do not adhere well to wallpaper. The same general rules for preparation that apply to uncovered walls apply here. The paper must be clean and fastened securely to the wall. Never paint over dry-strippable paper, since this can be removed with ease.

If the wallpaper has greasy stains on it, wash it with water and soap or detergent. Modern papers are washable, and grease can be removed without too much difficulty. Loose dirt and dust must also be removed. Wipe off excess water with a dry sponge.

If the paper is peeling or torn in several places, it should be removed before the wall is painted. Wallpaper removal is described in Section 8-4.

If there are only a few bubbles where the paper has pulled away from the wall, and the paper is otherwise sound, it is not necessary to strip it. With a sharp knife cut two slits in each bubble to form a cross. Lift up each of the four corners and apply some glue or paste to the back. Then press the paper back to the wall, wiping off any glue that squeezes through. When the glue dries, you can paint over the paper.

4-5. Preparing Wood

While wood floors require special treatment (discussed in Section 5-5), there are many other wooden surfaces in a house. These include window trim, doors, woodwork, panelling and furniture. The general rules for preparation apply: fill holes, sand smooth and clean off dust and grease.

Holes, scratches and dents can be filled with special wood fillers. If nailheads are visible or new nails are driven in, drive the heads below the surface with a nailset and fill the holes. Some fillers can be applied directly; with others, the wood must be primed first. Always read the instructions on the package. After the filler is dry, the surface should be sanded smooth and then wiped clean.

Knots in wood are a special problem. If they are loose, glue them in place and then sand them smooth. Because knots may exude resin which could bleed through the paint job, they should be sealed with a knot sealer or ordinary shellac. Sand the shellac lightly so that the paint will adhere to it better.

Defects should be removed from the wood. If there are burned areas, sand down to bare wood. If part of a board is broken, cut out the broken piece and replace it with a new piece cut to shape. Make sure nailheads are set below the surface and covered with filler. If the wooden surface has an old paint coating that is peeling, strip off the paint as explained in Section 4-3.

4-6. Preparing Metal

Metals may require careful surface preparation to make sure that the paint adheres well. The preparation depends to some extent on the type of metal and whether it has been painted before.

Ferrous metals (iron and steel) rust easily when exposed to dampness. If there is any trace of rust on the surface to be painted, remove it with steel wool or a file. Then clean with a suitable solvent. As soon as the metal is clean, apply a metal primer that inhibits rust. If bare steel is left exposed for even a short time, moisture in the air can start it rusting again. Water should not be used to clean the metal, for the same reason. Thus, mineral spirits, naphtha or a similar solvent must be used. Water-base paints should not be used on ferrous metals, but a latex metal primer is available for use under latex paint.

Aluminum also oxidizes in the air, but unlike rust, the oxide is hard and takes paint readily. Aluminum must be clean before it is painted. New aluminum usually has an oily film that can be washed off with warm water and soap or a solvent. Brass and bronze also tarnish in air, but like aluminum, paint adheres well to their oxides. If any of these metals has old lacquer on it, the lacquer should be removed before new lacquer or paint is applied. Use lacquer thinner to remove the lacquer, and wash it off with undiluted household ammonia.

If the metal surface is covered with an old paint film that is in good condition, new paint can be put on over it. Just make sure the old paint is clean. Wash off dirt and grease with trisodium phosphate or detergent and warm water.

4-7. Preparing Masonry

Masonry in the house, such as a fireplace or basement walls or floor, does not need paint for protection. However, you may decide to paint this type of surface for ornamentation or for ease in maintenance. Since latex masonry paint can be applied directly to masonry, you need only make certain that the surface is clean.

If a brick fireplace has loose mortar, it should be scraped free before painting. Then put in new mortar. Wash the surface and sweep up all dust before painting.

If masonry has an old paint finish on it, you can generally paint over it as long as the finish is in good condition. Rarely will you have to remove paint from a masonry surface. If there is some peeling, loosen as much of the old paint as possible with a wire brush and then feather the edges of the paint which remains.

4-8. Electric Fixtures and Hardware

Before you begin to paint a room you will have to decide whether you will paint over hardware and lighting fixtures or whether you will leave them unpainted. Usually, switch plates

Fig. 4-2. Removing electrical switch plate.

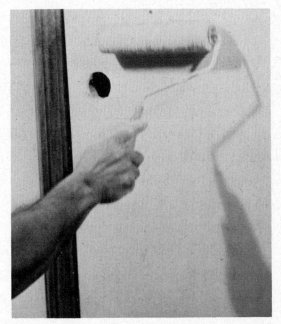

Fig. 4-3. Knob removed from door.

and outlet plates are painted the same color as the walls, but doorknobs and lighting fixtures are not painted.

Whether or not you are going to paint switch plates and outlet covers, you should remove them before painting, as shown in Figure 4-2. If you leave them on the wall and paint over them, some paint may seep behind them. When this paint dries, it will be impossible to remove the plates without damaging the surface. If you want to paint the plates, spread them on a pile of old newspapers and paint them when you paint the wall. To make sure you will remember to paint the heads of the screws(and incidentally to remember where the screws are), put them back in their holes after you remove the plates. When you paint the wall, you can touch the heads in passing.

Light fixtures are rarely moved, but again, if you try to paint close to them, some paint can seep in and stick the fixture tightly to the wall or ceiling. Then if you have to get at the electric circuit underneath, you may damage the wall. To avoid this, unscrew the fixture and let it hang by its wires. When the paint is dry, you can fasten it in place again.

Similarly, it is easier to paint doors if doorknobs, striker plates, hinges and locks

are removed first. With hardware removed, you can paint rapidly and more than make up the time and effort required to take off the parts and put them back. In Figure 4-3, a roller is used to paint a door with the doorknob removed. No cutting in around the knob is necessary.

Before you begin painting a room, you will, of course, take down curtains, draperies and all pictures on the walls. If the pictures are to be put back on the same hooks, no other preparation concerning them is required. You can paint the hooks the same color as the wall. If the pictures will be moved, pull the hooks and seal the holes before painting. The hooks can be put in their new positions before or after the walls are painted.

4-9. Protecting Floors, Furniture and Other Surfaces

The first step in getting ready to paint a room is to clear everything out. The time spent moving the furniture is small compared to the time you will save by not having to climb over and around it. Since it is not always possible to move everything out, compromise is often necessary. If you are going to paint the walls only, leaving the ceiling untouched, you can move all the furniture into the middle of the room, leaving a passageway next to each wall. If you are going to paint the ceiling, move the furniture to one end of the room and paint the ceiling at the other end. Reverse ends and do the other half of the ceiling. Finally, move the furniture to the middle and paint the walls.

If the furniture remains in the room, it must be protected with drop cloths. Flat surfaces like tables can be covered with newspapers. The floor must also be protected whether the furniture remains in the room or is taken out. Again, drop cloths or newspapers can be used.

When painting a wall close to the floor, you should use a paint shield to keep paint off the floor, but don't depend on the shield alone to protect the floor. Cover the floor with newspapers right up to the walls.

If you wish, you can use a paint shield in many tight places, such as on door panels and window sashes, and at junctions where two different colors of paint meet. If you want to avoid using a shield, stick masking tape on the surfaces to be protected before painting. After you finish painting, you pull off the tape and the excess paint with it. Masking tape is available in widths up to 4″. For most applications you can use a 1″ tape if you paint carefully. If you want to be freer with your brush, buy a 2″ tape.

4-10. Summary

Before you paint a room, you must get it ready for painting. The steps involved are:

1. clearing the way

2. repairing physical damage

3. preparing the surface to be painted

4. cleaning the room and the surfaces

5. protecting unpainted surfaces

5

Painting

After you have prepared the surface and moved the furniture, you are ready to begin painting. If you work conscientiously, you may expect to use somewhat less than a gallon of paint per person per day. A gallon of paint covers from 200 to 500 square feet, depending on the surface. Rough, porous surfaces require more paint. A second coat always requires less paint than a first. If you figure about 350 square feet per gallon and 350 square feet per day you will have a rough estimate of how much paint you will need and how long the job will take. When you order paint, buy one or two extra cans and arrange with the paint dealer to allow you to return unopened cans for full credit. This ensures that you won't run out of paint in the middle of the job.

5-1. How to Handle Paints

Paint is a mixture of pigment in a vehicle or carrier, such as oil or water. The pigment is not dissolved in the vehicle, but is held in suspension. When a can of paint remains unused for an extended period of time, the pigment settles to the bottom of the can. Before the paint can be used, it must be mixed thoroughly so that the pigment is put in suspension again. When paint is left in the can overnight, the pigment settles again, and the paint again must be mixed. If the paint is not mixed thoroughly, the color on the painted surface will vary, and the pigment will be distributed unevenly so that its protective property will vary from place to place.

When you buy paint, the dealer will mix it for you on a machine if you intend to use it within a few days. The mechanical shaker agitates the can so that the pigment and vehicle are thoroughly mixed. Alkyd and oil paints can be used immediately after shaking; latex paints should stand unopened about an hour before using. Never shake or stir varnish.

If you must mix paint yourself (because you want to finish a partially used can, for example), you can ensure a thorough blending by following these steps: pour the surface oil that has risen to the top into a clean, empty container such as another paint can or a cardboard mixing pail. Stir the remaining oil and pigment with a wooden paddle until they are thoroughly blended. Gradually pour oil back into the mixed paint, stirring as you do so. Finally, pour the mixed paint back and forth between the two pails several times.

Try to have enough mixed paint available to do the complete job, or at least a whole wall. Different cans of paint may have slight differences in color, even if they have the same batch number. If you change cans when you go from one wall to another, the color change will not be noticeable; but if you change in the middle of a wall, it will be.

If you add a thinner, as is usually required when you use a roller, make sure it is right for the paint. Read the label. Do not add too much thinner. Since the thinner causes a slight change in color, you must add exactly the same amount to each can of paint. Add a little at a time, stirring constantly. Finally, pour the mixture back and forth between two containers.

The paint can tends to get messy as you work. The groove that holds the lid usually gets filled with paint, and the outside of the can is frequently wet with paint. When you put the lid back on the can when you are finished, the paint in the groove can splash out and make a mess. One way of avoiding this is to punch holes in the bottom of the groove with a hammer and nail. Most of the paint in the groove will flow through the holes back into the can. The lid seals the holes when it is in place. To replace the lid, put it over the groove and cover the whole can with a cloth. Then tap on the lid with a hammer. Any paint that splatters will soil the cloth, but nothing else. Another way to avoid the mess is to crimp aluminum foil all around the top edge of the paint can. When you are through, discard the aluminum, and the paint mess will go with it, leaving the groove free of paint.

Old paint is usually lumpy and should be strained before it is used. Cheesecloth or an old nylon stocking can be used as a strainer. Place the cloth over a clean container, and pour the paint through it. Discard the cloth and the lumps.

When a paint job is to be continued the next day, the paint must be poured back into the can and the can sealed. Be careful about splashing paint when sealing. When the can is reopened, note if a film has formed on the paint. If so, remove and discard the film before stirring the paint. You should clean out the brushes and leave them in thinner overnight.

When you are painting, you will want to put newspaper under the can to catch drips. Or better yet, glue a paper plate to the bottom of the can.

When a job is completed, the equipment must be cleaned. If you used latex paint, simply wash everything in soap and water. Shake out all water from brushes or rollers and wrap them in newspaper to keep them free of dust. If you use turpentine or other thinners to clean your tools, pour the thinner into a shallow pan and put your tools in the pan also. Work the thinner into every part of the brush until all traces of pigment are out. Then wash your tools in soap and water to remove the thinner, comb out the brushes and wrap them in newspaper.

Save the dirty thinner. Pour it into a jar or a coffee can that can be sealed with a plastic cover. After a few days, the paint will settle and leave clear thinner that can be poured off and used again. Store the clean thinner in a suitably labeled container. If it is flammable, make sure you note that on the label also.

5-2. How to Paint a Room

The ceiling and walls of a room are painted from top to bottom. Paint the ceiling first, then the walls. Doors, windows and woodwork are done after the main areas of the wall are finished. Paint the baseboards last, since your shoes might rub against them while you are painting the wall. Before you begin painting, stir the paint, even if it was stirred by the paint dealer when you bought it.

Paint the ceiling with a roller or a brush. You will find the roller easier and faster, especially if it is equipped with an extension handle. If you paint in strips across the shorter dimension of the ceiling, you can paint each overlapping strip before the last edge dries. Latex paint is very good for hiding lap marks, but occasionally lap marks show up later if one strip is dry before the next is applied. Cut in at the corners and junctions of ceiling and walls with a brush or corner applicator.

The walls are painted with a brush or a roller, just as the ceiling was. When painting a wall, whether using a brush or a roller, a right-handed person should paint from right to left, and a left-handed person from left to right. In this way, if you accidentally contact the wall with your idle hand, you will be touching a section that has not yet been painted. You can also lean or brace yourself against the wall without any risk of marring the paint.

After you have painted the room, look carefully for any spots you have missed. Go over them with the same tool you used before. If you used a roller, touch up the spot with light, slow strokes of your roller. If you were brushing, cover the missed spots with the tip of your brush.

5-3. How to Paint Doors and Windows

When you paint a door or a window, you must remember 1) to paint every part and 2) to paint in such a way that wet paint on one part does not interfere with the painting of another part. Painting a window requires special care: in order not to have to lean over painted surfaces, it is preferable to paint the window frame after the window is finished.

Doors do not present the same problem as windows. You can paint the door frame either before or after you paint the door, at your own convenience. If a door is one unbroken surface, there is not much problem on deciding the order of painting. Remove the door-knobs and other hardware and paint the edges first with a brush. For interior doors, you can omit the bottom edge, since no one sees it, but for exterior doors, the bottom edge should be painted to protect the door from seeping moisture. Remove the door from the hinges and paint the bottom edge with any quick-drying coating. Then rehang the door and paint the other edges. Finally, paint the surface of the door with a roller or brush.

If a door has panels, there is a preferred order of painting the parts to minimize the possibility of lap marks. First, the edges are painted; next the moldings around each panel (both of these are done with a brush, but the rest can be done with brush or roller); third, the panel; fourth, the horizontal members or rails; and finally, the vertical members or stiles. After painting, leave the door ajar until the paint has dried. Then replace the door-knobs and other hardware.

The order of painting the parts of a double-hung window is as follows (all parts are painted with a brush): first, raise the lower sash as high as it will go and lower the upper sash part way. Now begin by painting the check rail on the top sash. This is the bottom horizontal member. Second, paint the horizontal and vertical bars that divide the sash into small panes. Some windows have such bars in one or both sashes, while others may have one large pane in each sash. The vertical members (stiles) of the sashes are painted next. Throughout the foregoing procedures, it is necessary to raise and lower the sashes to get at the surfaces. You can grasp the horizontal rails of the sashes to do this, since they are painted last.

When painting a window, you have to be careful to keep the paint off the glass. You can mask the window panes before painting or use a shield; or you can just try to be careful. Don't worry too much if you get a little paint on the glass. Let it dry, and then remove it with a razor blade. If a door has a glass pane in it, paint the molding around the pane before painting the rest of the door. Use the same precautions to keep paint off the glass, or scrape it off later, after it dries.

5-4. How to Paint a Floor

Interior floors are usually finished with a transparent coating, as described in the next section, so that the natural grain of the wood shows through. Paint is opaque and hides the beauty of the wood. On the other hand, paint also hides any flaws and blemishes in the floor. Outdoor floors, on decks and porches, for example, are usually painted, and some low-quality wood floors are also painted. Paint is sometimes used on concrete floors.

Any good paint will protect a floor, but

specially designed floor paints will resist wear longer. The simplest to use is latex floor paint. You can apply this paint over a damp surface since it is water-solvent. It dries in about an hour. However, you should wait until the next day to apply the second coat. Two coats are usually sufficient, except on bare wood, which requires a third coat. You can apply the paint with a roller or a wide brush.

Oil-based and alkyd floor enamels are the most common floor paints; they are more durable than latex. However, the cleaning-up process is more involved. These paints cannot be put down on a damp surface.

Rubber-based paints are best for concrete floors, although the other floor paints are also quite satisfactory. Rubber-based paints should not be used on wood floors. Since gasoline attacks rubber-based paints, do not use them on a garage floor. Rubber-based paints cannot be applied with a roller because they dry too rapidly. Although these paints are extremely durable on concrete and masonry surfaces, the average home owner should not use them because they are difficult to apply and require special solvents to remove them from the equipment after the job.

Latex or alkyd floor paints can be applied with a roller or a brush. Latex paints are usually applied as they come from the can, but alkyd paint may be thinned with turpentine. Follow the instructions on the can. A simple and effective method of painting concrete floors is to pour some paint directly on the floor and spread it with a stiff push broom.

5-5. How to Refinish Wood Floors

When a floor has a few scratches in it, you can usually repair the damage without completely refinishing the surface. If the scratches do not penetrate further than the finish, they can be removed with steel wool. If there are stains as well as scratches, use a cleaning agent with the steel wool. Rub only in the direction of the grain. When the scratched finish is removed, smooth the surface with fine sandpaper or fine steel wool and apply a matching shellac or varnish. Dilute shellac slightly with alcohol; dilute varnish with turpentine.

If the scratches are deep and penetrate the floor, remove the finish, as indicated above, and smooth the wood. Then fill the scratches with wood filler or wood plastic to a level above the surface of the floor. Remove excess with a putty knife. When the filler is dry, sand it down flush with the floor with smooth sandpaper. Finish with shellac or varnish, as described above.

Refinishing a floor completely is not a difficult job, but it requires special equipment which can usually be rented. There are three steps in refinishing: (1) sanding off the old finish; (2) applying a finishing coat of varnish or shellac; and (3) applying a coat of wax to protect the finish.

Before beginning the sanding operation, remove all furniture from the room, including pictures, drapes and Venetian blinds. Sanding leaves a thin layer of dust on everything. Look carefully over the entire floor, and if you see any nailheads, drive them down with a nailset. This is a good time also to nail down any loose boards or squeaky spots in the floor. Open the windows, shut the doors, and you are ready to begin.

You will need to rent a drum sander and a disc sander. The drum sander is used first to cover all large areas in the room. Sand parallel to the grain. Follow the directions supplied with the machine, and you will find it amazingly simple to operate. The disc sander is for edges which are not accessible to the larger drum sander. With both sanders, begin with a coarse sandpaper to remove the finish, then progress to finer papers for a second and third sanding.

After completing the sanding and before beginning the finishing, clean the whole room, including tops of doors, windows and baseboards, with a vacuum cleaner. Any dust left in the room can spoil the finish.

Finishes for floors fall into three categories. Floor sealers soak down into the wood and do not form a film or coating on the surface. Varnishes, shellacs and some synthetic finishes form a surface coating which may be colored but is usually transparent so that the grain of the wood is visible. Floor paints cover the floor with an opaque coating. The choice depends on the type of finish desired. Sealers and transparent finishes are covered in this section; painting a floor is discussed in the preceding section.

If you want a matte finish on the floor, you should use a floor sealer. Apply the first coat with a brush or mop, or wipe it on with a cloth. If you wish to add color to the floor you can get colored sealer. Let the sealer remain on the floor for about fifteen or twenty minutes, and then wipe off the excess. Let it dry overnight, then rent an electric buffer and buff the floor. Now add a second coat of sealer in the same way. The sealer finish is usually scratch-proof. Worn areas can be touched up simply by applying more sealer, with no danger of lap marks showing.

Varnish makes a pleasing finish, but it is slow-drying and therefore awkward to use. So-called "quick-drying" varnishes should not be used since they are not durable. Varnish can be put down over an old finish, if the old finish is in good condition. For new or resanded wood, use a sealer for the first coat. After it has dried overnight, apply the varnish. You should put on at least two coats of varnish, waiting at least twenty-four hours for each coat to dry. Alternatively, you can use a coat of thinned shellac as a sealer, and then two coats of varnish. Allow each coat to dry thoroughly before the next is applied. A varnish finish darkens with age.

Shellac is simple to apply and dries quickly. White shellac does not stain the wood, so that the natural grain is visible. Orange shellac also shows the grain, but darkens the wood somewhat. Shellac should be applied in thin coats. Shellac as purchased should be thinned with alcohol. Ask your paint dealer how much to thin it, since dealers may sell their own shellac already thinned. For bare wood, plan to put on three coats of shellac, allowing each coat to dry at least two hours before putting on the next. Shellac is not as water-resistant as varnish, but it is easier to apply, and worn areas can be touched up easily.

Synthetic or plastic finishes are the most durable. They *require no wax* and will last for years. They must be thinned with a special thinner and can then be brushed on easily. Two coats are sufficient. These materials are available in either a glossy or matte finish.

When your final finish is dry, you should wax the floor to protect the finish. Do not use a self-polishing liquid wax on a wood floor, since these waxes contain too much water, and can damage wood. Paste waxes and solvent-based liquid waxes (not self-polishing) are satisfactory. Apply with a soft cloth and allow about twenty or thirty minutes to dry. Then use the electric buffer to polish. Two thin coats, buffed separately, give a longer lasting shine than one thick coat.

5-6. How to Paint Metal

Metal surfaces such as air registers or frames of casement windows are no problem if they have once been painted. Simply clean the surface, as explained in Section 4-6. If the paint is not peeling, just paint over it with a brush, as shown in Figure 5-4. If you are painting a wall with a roller, paint any air vent in the wall with a brush at the same time you use the brush to cut in the corners.

Latex paint can be applied directly to bare metals. However, since ferrous metals can rust when exposed to moisture, they should be primed before the latex paint is applied. Use special latex metal primers before painting with latex paint.

Alkyd and oil paints cannot be applied on bare metals without primers. Zinc chromate is a good primer, but be careful when using it, since it is poisonous. Let it dry overnight and then apply the finishing coat.

5-7. How to Paint Masonry

There are many different kinds of masonry

Fig. 5-4. Brushing paint on metal grill.

paints, and they are not usually compatible. If you are going to use the same type of paint that is on the surface already, it is sufficient to clean the surface of dirt and loose paint and simply paint over it. If the new paint is different from the old, or if you do not know what kind of paint was used originally, you will have to use a primer to make sure the new paint sticks.

Latex masonry paints are the easiest to use because they are thinned with water. Latex paint dries quickly and is very durable, but it cannot be used over old layers of alkyd or oil-base paint. However, there are latex primers available which adhere to other paints and also make a good base for latex paint.

5-8. How to Finish Furniture

If you build furniture or buy ready-to-paint furniture, you can save money by finishing it yourself. The term "finishing", when applied to furniture, means covering the object with paint, varnish, lacquer or another film to protect its surface as well as to enhance its appearance. A piece of ready-to-paint furniture is about 20 to 30 per cent cheaper than the same piece finished. Finishing, however, with the exception of painting, is a long, slow process. Although each coat may be put on

quickly, the surface should be allowed to dry for at least twenty-four hours between coats; at least four coats are usually required. In addition, the surface must be sanded or rubbed several times during the process. You must decide whether the money you save by buying ready-to-paint pieces justifies the work.

If a piece of furniture is functional rather than decorative, such as for use in a kitchen or playroom, you may decide that it need only be painted. Although paint covers the wood completely so that the grain cannot be seen, it also covers blemishes and permits you to build furniture out of odd scraps of wood with grains or patterns that don't match. This would be impossible if you wanted a transparent finish such as varnish, shellac or lacquer. The transparent finishes are used when you want the natural beauty of the wood to be visible through the finish.

Before beginning any finishing job, plan your work. Remove all hardware, such as locks, metal knobs and handles. Drawers should be removed from chests or cabinets and placed with the front surface, the surface to be painted, in a horizontal position. Plan to coat the inside corners and hard-to-reach areas first. Save the easily accessible surfaces for last.

Painting is very simple. Any good enamel can be used. First, the surface must be dusted thoroughly. Then, use a 2″ brush to apply an undercoat to the bare wood. This is usually a white enamel undercoat compatible with the final coat, but it can also be tinted. If you use latex enamel, the same enamel can be used for both undercoat and final coat. Only these two coats are necessary. The undercoat should be allowed to dry for at least twenty-four hours. Then the surface is sanded with fine sandpaper until it is smooth. Dust the surface thoroughly to remove all traces of dirt and grit. Then apply the final coat. After another twenty-four hours the piece of furniture is ready for use. Paint should not be used on expensive, beautiful woods since it hides the natural beauty.

Varnish is difficult to apply. The biggest problem is air bubbles which appear on the surface and prevent the formation of a smooth finish. There are various ways to minimize air bubbles. To begin with, never shake varnish; if you stir it, do so very slowly. Dip the brush in varnish so that only one-third of the length of the bristles is below the surface. Instead of wiping the excess off the brush by drawing it across the lip of the can, tap it against the edge. Apply only light pressure to the surface, rather than pressing down too hard. Allow the varnish to flow on with long strokes. There are special brushes for varnishing that are very soft and have many flagged bristles to hold more varnish and fewer air bubbles.

The first coat of varnish may be thinned with one part turpentine to nine or ten parts varnish. Succeeding coats are not thinned. Each coat should be allowed to dry for at least twenty-four hours and should then be sanded smooth with very fine sandpaper. When sanding, always rub with the grain so that minute sanding marks do not show. Remove all dust before and after each sanding, and especially before applying the next coat. At one time, the final coat needed to be rubbed by hand to produce a satin finish rather than what was considered an objectionable mirror-like surface. Now, semi-gloss and flat varnishes that need no rubbing are available for the final coat. The glossy varnishes are tougher and are used for the first two or three coats; then a "satiny" varnish is used for the top coat.

If you want something to protect the wood and are not too particular about appearance, apply one or two coats of shellac. Shellac may also be used on good woods, since the grain shows through. For a fine finish on dark woods, you can apply about four coats of shellac, sanding each coat with very fine sandpaper. Shellac is not water-resistant.

Stain is used to color wood without hiding the grain. It is easy to apply. On soft wood, however, stain soaks into the ends more than into the flat surfaces, causing the ends to be darker. This can be prevented by first coating the ends with a thin layer of shellac. Stains can be brushed on easily since they do not show brush marks.

When the wood is stained to the desired tone, the stain finish can be protected by covering it with clear lacquer. The only problem with applying lacquer is that it dries very fast, so you must work as fast as possible without retracing your steps. Apply in long strokes from one side of the surface to the other, and make each stroke slightly overlap the preceding one. One problem with lacquer is that since it is clear, you can't see whether you've missed a spot until it has dried. You can go back then and touch up the areas you missed. You should apply at least three coats of lacquer, although five is preferable. After each coat (except the last) is dry, rub the surface smooth with a ball of very fine steel wool. Dust the surface before applying the next coat.

Refinishing furniture involves removing the old finish and then finishing by one of the methods described above. If you plan to use paint, however, it is not necessary to remove the old finish as long as it is hard. In fact, you may be able to get by with only one coat of enamel, since the old finish can act as an undercoat.

Removing the old finish is accomplished easily with chemical removers. Buy one that is non-toxic and non-flammable. The general procedure for using these chemicals is explained in Section 4-3, but always follow the instructions on the package. After the old finish is removed, wash the wood with ordinary water to remove all traces of the chemical. When the wood is dry, you are ready to begin refinishing.

Before you start refinishing, make sure the surface is in good condition. Cover all scratches; fill screw holes, knotholes, or cracks, with a wood filler (the filler can be stained to match the wood). Light scratches can be removed by rubbing them with furniture polish thinned with rubbing alcohol. Scratches in walnut can be stained by rubbing them with a piece of walnut meat. Scratches in mahogany can be touched up with iodine. Matching oil stains for most woods are available. After the surface is repaired and dusted, finish as desired.

Kits are available for special finishes for furniture. Plastic and wood veneer laminates with adhesive backing can be applied directly to surfaces, even to surfaces built up out of scraps of wood. Antiquing kits enable you to make a new piece of furniture look like a genuine antique. Gold leaf and other special effects are all easily applied with kits on sale in building supply stores and in mail-order houses.

5-9. Safety

Before, during and after painting, there are certain safety precautions which must be taken. The principal dangers are falls from ladders, chemical irritation, poisoning and fire. At this point you may think that the risks of doing the job yourself outweigh the potential economic gain, but actually there is little danger so long as you are aware of the hazards and take simple precautions.

For painting a ceiling or upper wall, use a stepladder rather than a box or a chair; it is a common practice also to place a board across two chairs so that you can paint a large area without having to go up and down too many times. The procedure is safe as long as the board is solid enough and wide enough, and you don't lean too far.

Paint fumes may be toxic, especially if the paint contains lead or mercury. To minimize the dangers, always provide as much ventilation as possible when you are mixing paints, painting, and cleaning equipment afterward. When removing old paint with a scraper or blowtorch, paint dust may get into the air; the dust can be very toxic if the paint contains lead. It is advisable to use a respirator when removing paint with a scraper or torch. This danger does not exist when a chemical paint remover is used.

Paints and thinners can irritate the skin on contact. If you get solvent on your hands, wash it off immediately. Better yet, wear plastic gloves when handling turpentine, naphtha, other solvents or any chemical paint remover. Rubber gloves may also be used with most chemicals, but not with solvents for rubber-based paints. Plastic gloves are cheap and can be thrown away after each use.

Avoid the risk of fire by using non-flammable thinners and cleaners wherever possible. If a flammable substance is used, dispose of rags in a tightly sealed metal container. Do not leave oily rags lying about as they can ignite by spontaneous combustion.

When pouring paints or thinners, take care that there is no splashing which can irritate the skin or get in your eyes. It is not necessary to wear safety glasses when painting, but do take care not to touch your eyes or get anything in them.

Many of these precautions are unnecessary when you use water-soluble paints. You can also avoid the dangers of lead poisoning by using non-toxic paints. On children's toys and the walls of children's rooms you should always use lead-free paints.

Flexible Wallcoverings: Materials and Designs

Wall decorations date back to the cave man. Medieval wallcoverings served as insulation against cold and dampness as well as being decorative. They were usually woven and it was a small step to include a design to enhance the appearance. In the homes of the wealthy, rich tapestries served both purposes. Wallpaper was originally conceived as an inexpensive decorative material for those who could not afford more expensive tapestries or patterned textiles. The wallcoverings that are pliable, like paper, are called *flexible,* to distinguish them from *rigid* wallcoverings such as wallboard and panelling.

6-1. Materials

Though wallcoverings go back centuries, technical advances are comparatively recent. Colors that did not fade were developed in the 1920's and washable colors in the thirties. These two developments opened up new possibilities in design. Silk screen printing, in the forties, was a step toward mass production. The fifties saw the development of prepasted and pretrimmed paper, making it easier for the home handyman to hang wallpaper without the aid of a professional. The biggest development in the sixties was the widespread use of synthetic materials both as a base and as a coating for paper and cloth. Synthetic coatings led to the development of *strippable* wallpaper that could be pulled off a wall without soaking or scraping. This was a boon to the handyman, for if he made an error in hanging, he could strip off the paper and reapply it correctly. When he got tired of the paper and wanted a new design, he could pull off the strippable paper without effort.

Vinyl is a synthetic material used in the manufacture of some wallcoverings. It is available either as a thin film or a liquid. In liquid form it may be applied to a backing of paper or cloth to form a *vinyl-coated* covering. The term *vinyl wallcovering* is used to describe coverings of laminated vinyl and other materials. Vinyl may be laminated to paper, cloth (either natural or synthetic), or other synthetic bases. A laminate of paper and cloth which is coated with vinyl may be referred to as vinyl wallcovering. Dry-strippable coverings usually contain vinyl.

Flocked wallcoverings feel like silk or velvet. To flock a wallcovering, the design is printed in varnish, shellac, or some similar sticky substance. Finely shredded fibers are then spread over the covering, and adhere to the varnish. The surface can also be coated to make the wallcovering washable.

Wallpaper designs are not printed to the edge of the roll, so that a white border remains on each edge. This is called a *selvage,* and instructions for matching or hanging were sometimes printed there. The selvages also protect the edges of the papers during shipment and handling. The paperhanger first has to trim off the selvage before beginning the job. Although it is possible to hang strips of wallcovering with edges overlapping, so that only one selvage need be removed, the preferred method is a *butt* seam with no double thicknesses. Trimming selvages in a straight line to make an invisible butt seam is difficult, and to simplify the work as much as possible, manufacturers offer *pretrimmed* wallcoverings, with selvages removed at the factory. They are well worth the increased cost and extra care in handling.

Another improvement to make wallpapering easier is *prepasted* wallcovering. Like postage stamps, the coverings are backed with an adhesive which is activated when dipped in water. Instructions for hanging prepasted coverings are included with the materials, including the temperature of the water and the length of time to dip. To simplify the hanging procedure further, inexpensive water containers are available to be used at the base of the wall being covered. For those who prefer to apply paste themselves, ready-mixed adhesive is available that is applied just as it comes from the can.

A roll or *bolt* of wallcovering is a standard unit containing 36 square feet, no matter how wide the strip. A double roll contains twice this (72 square feet). In practice, you should expect a roll to cover about 30 square feet, since there is always some waste. A double roll will cover more than 60 square feet, since waste is not doubled.

6-2. Designs

There is no limit to the number of different designs that can be executed in wallcoverings. In fact, any imaginable design can be duplicated, and most have been in existing wallcoverings. Once, wallcovering designs came in about half a dozen categories — florals, geometrics, scenics, and the like — but it is no longer possible to limit even the categories since new ones are devised regularly.

Most wallcoverings are patterned. That is, the design is repeated regularly, usually on a single strip. The distance from the center of one element of a pattern to the center of the next is called a *repeat.* The distance is usually measured horizontally and vertically and the two readings are called the *horizontal repeat* and the *vertical repeat* of the pattern.

Scenics or murals are wall decorations with pictorial designs that occupy two or more strips. A true mural does not repeat itself, but in wallcoverings a mural may cover most of a wall and then be repeated around the room. The picture in a scenic wallcovering may be anything, real or imagined. Pictures used include battle scenes, historical incidents, mountains, lakes, rural scenes, statuary, nudes and people.

Florals are probably the most common wallcovering patterns. They may depict flowers, leaves, or plants, from realistic representations to primitive imitations. Geometrics are designs that feature geometric shapes. These include stripes, polka dots, plaids, checks, and lattices. Geometrics are best identified when they are called by what they show, such as a polka dot pattern, or harlequin pattern.

Wallcoverings can be made to imitate other materials, including designs used in cloth. Thus, there are damasks, paisleys, corduroys, and others that depict textile designs. Also available are coverings that imitate wood panelling, basketweave, grasscloth, marble, and in fact, any other material that one could conceivably want on a wall.

Wallcoverings can be used to depict historic events in repetitive patterns and imitate the art of different periods. Thus, there are oriental motifs, Renaissance designs, and psychedelic, pop art, and op art designs. Georgian is reminiscent of the style popular in England in the eighteenth century; French Provincial represents the rustic style of the French provinces in the same period. Other period designs include Louis XV, Louis XVI, Regency, and Jacobean.

Some designs copy those used for other purposes. Heraldic coverings show motifs representing crests and coats-of-arms. Tea chest paper designs depict the small geometric patterns used in the orient on paper in which tea is packaged. Bandbox designs show the kinds of motifs on bandboxes and hatboxes in the early nineteenth century.

Ceiling papers are coverings with plain geometric patterns that can be viewed from any direction without looking as if they are upside down. Ceiling papers can be used for special effects. The strips can be cut in triangular shapes and hung so that the points are at the center of the ceiling. This is called *canopy ceiling* design. If the ceiling in a room is high, it can be made to appear lower by bringing the ceiling paper down the walls to a border or molding. This is a *drop ceiling* design.

There are wallcovering designs for specific rooms. Kitchen paper may have a design featuring baskets of fruit or kitchen utensils. Nursery characters are usually represented on wallcoverings for children's bedrooms. Coverings for a playroom include designs featuring playing cards, monopoly layouts, and dice.

Conversation piece designs have to be chosen carefully. They are intended to be so unusual that they immediately draw attention from the rest of the room. But their novelty can wear off if used in a prominent place. However, in a quiet room such as a bedroom, a conversation piece can be very effective.

Manufacturers often put out designs in sets of two or more which are compatible for adjoining rooms or separate walls of the same room. For example, a simple rattan pattern may be used on one wall, and the same design with the addition of small flowers, twigs, or some other accent can be used on a second. The plain rattan design can be used on three walls, and the accented design on the fourth. Designs in adjoining rooms should be different but should not clash.

Most designs come in a variety of colors, so that you can usually find a design you like in a color to match the rest of your furnishings. For example, a simple floral may depict small white daisies on a solid color background. It would be available with a dark blue or dark green background for large rooms or with a pastel background for small ones. If you like a design but the sample clashes with colors in your rug or furniture, ask to see the same design in other colors.

6-3. Names for Wallcoverings

Manufacturers number their designs, but also use names to make identification easier. Girls' names are often used, especially for florals, e.g. "Clarissa", "Melissa", and "Jane". Some names are simply descriptive, such as "Brick", "Lattice", "Denim", and "Gourmet". The last depicts foods and is a kitchen wallcovering. But manufacturers like to be distinctive, so we also have "Gourmet's Paradise". Many manufacturers have several different striped patterns, from wide to pin stripes, multi-colored and plain. What do you call a striped wallcovering? One maker came up with "Don't Let The Rain Come Down". Others are "Cumberland", "Antoinette", "Decorama", "Pencils", "Candy Stick", and "Lollipops". Some geometric designs are "On the Square", "Superstar", "Parquet", and "Amazin" (which is maze-like).

Choosing a Wallcovering

When you decide to paint a room, you have only to choose a color. But if you decide to use wallcoverings, you must choose colors or combinations of colors, then select from a vast number of designs. Don't be overwhelmed; with a little thought and planning, you can take advantage of this wealth of decorative possibilities.

7-1. Paint vs. Wallcoverings

By painting all the inside walls of your home you can end up with the monotonous look of a hospital ward. But if every wall were covered with a fancy design, the result might be too giddy for comfortable living. Before considering the question of paint or paper, decide whether a room should have designs or plain walls.

Generally, busy rooms should have plain walls. Rooms with natural decorations like bookshelves, a fireplace, or a picture window, should have walls that do not compete for attention with the built-in decorations. Similarly, rooms in which you plan to hang wall decorations do not need added design on the walls.

Rooms in which you spend little time can have interesting wallcoverings. Thus, a bed-room can have conversation piece designs that might be tedious in a living room. Rooms for routine tasks, kitchens, bathrooms and laundry rooms, can have designs to make your work more interesting.

Note that *design versus plain* is not the same as *paper versus paint* . There are wall coverings with very plain patterns that can be used wherever paint would be satisfactory. However, if a design is indicated or would be preferred for decorative purposes, then a wallcovering should be used. If a wall should be plain, your choice of paint or paper will be affected by factors of cost, convenience, and experience in painting or papering. Although you can buy very expensive wallcoverings with custom designs, it is generally cheaper to paper than to paint. Good quality vinyl coverings with plain patterns have a lower initial cost than a good paint job and will wear much longer than a coat of paint. Painting is somewhat easier than hanging wallcovering, but papering has no mysteries, and men and women who have never papered walls before find they can hang wallcoverings proficiently merely after reading instructions. Wallcovering tools are cheaper than painting tools, although most home owners already own paintbrushes and rollers.

In summary, you can find a wallcovering suitable for any wall in any room. You may elect to use paint where a plain pattern is indicated, but you should use wallcoverings

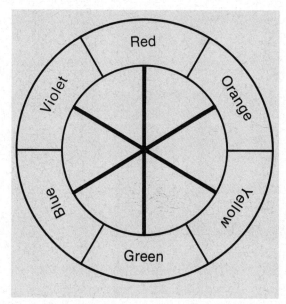

Fig. 7-1. Color wheel.

wherever a design can make a room more interesting.

7-2. Color Theory

White light is a mixture of all the colors of the spectrum. A rainbow occurs when sunlight is refracted through raindrops so that the different component colors are bent different amounts. When white light falls on a colored surface, a red wall for example, the pigments in the coloring on the wall absorb all the colors but red. Only red light is reflected back to our eyes. If the same wall were illuminated with a pure blue or green light, the light would be absorbed, since the pigments absorb everything but red. Then since nothing was reflected from the wall, it would look black. Black is an absence of color.

Pigments have three primary colors: red, yellow, and blue. All other colors, including white, gray, and brown, can be formed by a proper blending of these three primary colors. The color wheel in Figure 7-1 shows the relationships between the colors. The primary colors are red, yellow and blue. The other colors are called ''secondary''. Each secondary color can be formed by mixing the two primary colors on each side of it. Thus, orange results when red and yellow are mixed, green from a mixture of yellow and blue, and violet from red and blue. Notice that the color relationships are continuous around the circle. White is a blend of equal parts of the three primary colors, while browns utilize the three primaries in unequal proportions.

Two adjacent colors on the color wheel may be blended to form a new color. These tertiary colors are usually referred to by the names of the colors forming the new blend. For example, red-orange is a combination of red and orange, and blue-green is a combination of blue and green. In designating the tertiary colors, the primary color is named first. Decorators may use other names for some colors, such as tangerine for red-orange, or aquamarine for blue-green, but in choosing colors for color schemes (see next section), the source of the color should be kept in mind. Tertiary colors are essentially the result of unequal mixtures of two primaries. Thus, red-orange is a blend of red and orange, but orange is a mixture of red and yellow. So red-orange is a blend of red and yellow, with the red exceeding the yellow.

Opposite colors are called *complements* or *complementary colors.* Thus, red and green are complements, as are orange and blue, and yellow and violet. When complements are mixed, the result is a shade of white or gray. This can be understood from the color wheel. Since green is a blend of yellow and blue, then a mixture of green and red is really a mixture of red, yellow, and blue, the three primary colors.

From a practical standpoint, no pigment is *monochromatic.* A monochromatic color is a pure color. Most colors that look like a primary color may in fact be a blend of all three primaries with more than 50 percent of the dominant color.

7-3. Properties of Colors

Colors affect mood, can make a room seem

larger or smaller, and can accentuate certain features of furnishings. In a sense anything goes, and if you like a particular color combination, there is no reason why you shouldn't have it. However, if you know what can be accomplished with color, you may be able to make a room more comfortable or more interesting instead of simply acceptable.

Colors may be classified as warm or cool. Greens, blues and violets, and combinations such as blue- green and blue-violet, are cool colors. Reds, oranges, yellows, and their combinations, are warm. As the name implies, warm colors are stimulating and seem to advance. Cool colors are relaxing and seem to recede. Whites, blacks, grays and browns are neutral.

Colors may be light or dark. Objects painted in light colors appear larger, while those in dark colors seem smaller. Yellow makes objects appear largest. If a small room is painted a light cool color, the room will seem larger since cool colors make the walls appear to recede and light colors make them larger. Similarly, a dark, warm color can make a large, empty room seem cozy.

The use of color to change the apparent shape of a room can be startling. A high ceiling can be made to appear lower by making it darker than the walls. Likewise, a low ceiling can be "raised" by painting it white and using darker colors on the walls. A square, characterless room can be made more interesting by making one pair of opposite walls darker than the other pair. The room will then appear rectangular. Alternatively, one wall can be made brighter than the others, and it will act as a focus of attention.

A predominant color on the walls of a room will enhance the visibility of everything else of that color in the room. Thus, to accentuate a rug, piece of furniture, or painting, make a large area of the walls the same color as the predominant color in the featured object. Too many different colors in a room make it featureless. Do not use different colors in equal proportion. One color should always be dominant.

Cool rooms can be "warmed" by using warm colors. Rooms on the sunny side of a house should be in cool colors. Most rooms, however, should have warm and cool colors. Don't use all neutral colors to avoid conflicts. Without some warm or cool colors a room is dull. However, some neutrals should be used to contrast with the other colors in the room.

There are four basic color schemes that are usually used to decorate rooms, but you should feel free to use your own imagination and plan a room to your liking. The four schemes, however, are a good starting point and should be studied before letting your imagination roam.

A *monochromatic* scheme is built around one color. Walls, furniture, drapes, bedspreads are different shades or tints of the same color. This can be very effective in a bedroom, but may pall if viewed every day in a living room. If each bedroom is done in a different color, the whole arrangement can be very harmonious.

An *analogous* or *related* color scheme is very popular and can be used in any room in the house. Two or three colors that are close to each other on the color wheel are the basis of the scheme, with accents furnished by small tinges of adjacent colors. Thus, green, yellow-green, and yellow may provide the predominant motif with accents of blues or oranges.

Contrasting or *complementary* color schemes are also popular. Here you must be careful. Red and green may look like perpetual Christmas decorations, but by combining unexpected tints of complementary colors, such as pink and dark green, or pastel shades of both, the impact of the contrast is softened.

An *accented* color scheme is a combination of the analogous and the contrasting schemes. The predominant areas are covered with adjacent colors and the whole is boldly accented with a color from the opposite side of the color wheel.

7-4. Design Ideas

As with colors, designs can also change the apparent size and shape of a room. A low,

Fig. 7-2. Formal living room.

Fig. 7-3. Living room with floral pattern.

Fig. 7-4. Living room with geometric pattern.

cramped ceiling can be raised by using strong vertical designs, such as stripes, on the walls. The maximum effect is obtained when the ceiling is lighter than the walls, as described in Section 7-3. To "lower" a high ceiling, call attention to it with a pattern, or run a ceiling paper down the wall to a border or molding. Use horizontal patterns on the walls.

Cool colors on the walls make a room seem larger. Combined with open patterns or scenics, the effect of spaciousness is enhanced. On the other hand, big patterns with plenty of bright colors can make a bare room seem well furnished.

A narrow room should have horizontal patterns on the short walls. The horizontal lines will make the other two walls seem further apart. Do not combine horizontal lines on one wall with vertical on the next as this is confusing. Use floral patterns on the long walls.

Design and colors go together to create an impression. The colors and design of a wallcovering must not only be harmonious but must also blend with the furniture and decorations, especially with rugs, drapes, and upholstery. Further, designs in each room should be related to each other so that there is no "shock" of a sudden change of mood or impression.

The living room in Figure 7-2 contains light-colored furniture and rug. A light-colored wallcovering might make the other colors look washed out, so a dark green covering was chosen to contrast with and bring out the warm yellows in the rest of the room. The upholstered chairs have a matching green accent which is strengthened by the wall color. The small white bouquets on the wall break up the green expanse to give a more formal appearance than a solid color would have.

The living room shown in Figure 7-3 is a city living room, and the decorator wanted to create a mood of country intimacy in the city. Large floral patterns in autumn colors produce a feeling of warmth and snugness.

The living room in Figure 7-4 has a geomet-

Fig. 7-5. Kitchen.

Fig. 7-6. Bathroom.

Fig. 7-7. Penthouse.

Fig. 7-8. Boy's room.

Fig. 7-9. Dining room.

ric pattern in the sofa that is augmented by a different geometric pattern on the walls. The wall design repeats the blues and greens in several of the upholstery fabrics, calling attention to the Chippendale furniture without making the room seem cluttered.

The wallcovering shown in Figure 7-5 is typical of many bright, cheerful kitchen designs showing food or kitchen utensils. In general, a kitchen or any work area should have light walls for an illusion of roominess and warm colors for a cheerful mood. It is not always necessary to choose a design that says, "this is the kitchen". The most important property of a good kitchen wallcovering is that it be washable. The covering shown in the figure can be washed clean with a damp sponge and will outlast a paint job.

The design of Figure 7-6 is a floral that can be used anywhere. Here it is used in a bathroom on walls and vanity doors, and colors of towels and shower curtains are repeated in the wallcovering. Like kitchen wallcoverings, those in bathrooms must withstand moisture and be washable.

A geometric design was needed for the penthouse in Figure 7-7 to complement the contemporary setting of glass, steel and plastics. The intricate and intriguing design lends interest and warmth to what could be a very cold, functional room.

The two wallcovering designs in Figure 7-8 show how striking patterns can dress up a room to create a feeling of luxury with inexpensive furniture. An important aspect of covering the walls of a nursery or a child's bedroom is to choose a design that will not seem too childish as the children grow older.

A floral is always a good design for a dining room, as shown in Figure 7-9. Here the background color of the wallcovering matches and accents that of the chair seats. In dining rooms the wall design should not be so bold as to compete with the food for attention.

The wallcoverings in Figure 7-10 illustrate

Fig. 7-10. Apartment.

how good design can "fill" an empty apartment. The jungle design in the dining room blends well with the colorful cloth covering a cheap, secondhand table. The rattan design in the far room is related to the tropical motif of the dining room pattern and the two present a unified decoration.

The "mood" of a room may be determined by the wallcovering and should match the temperament or personality of the person who will live there. The living rooms depicted in Figures 7-2, 7-3, and 7-4 convey different feelings of warmth, intimacy, or cool formality. Make sure the design you choose is one you will be comfortable with.

How to Hang Wallcoverings

Manufacturers of wallcoverings have added so many features for the handyman that hanging wallpaper is now a simple task. This chapter covers the following steps in the process:

1. Measuring the room to determine how much to buy

2. Tools needed

3. Preparing the wall

4. Removing old paper

5. Hanging the wallcovering

6. Hanging prepasted wallcoverings

The procedure is not difficult, and most people who buy wallcoverings do their own hanging.

8-1. Measuring

A roll of wallcovering contains 36 square feet of usable material, regardless of the width of the roll. When you buy a bolt of material, it may be a single roll, a double roll, containing 72 square feet, or a triple roll (108 square feet). When you hang wallcoverings and have to match the pattern from one strip to the next there is always some waste. Taking waste into consideration, a roll of material should cover about 30 square feet instead of 36. If the design is one that can be matched at random, there will be very little waste. There is less waste with wider rolls since fewer strips have to be matched.

To find the area of the walls, first measure the perimeter of the room and multiply this by the height. Assume your room has the floor plan shown in Figure 8-1. The perimeter is $14 + 16 + 8 + 5 + 6 + 21 = 70$ feet. If the walls are to be covered from floor to ceiling, then multiply 70 by the height of the room to get

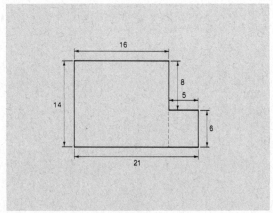

Fig. 8-1. Room dimensions.

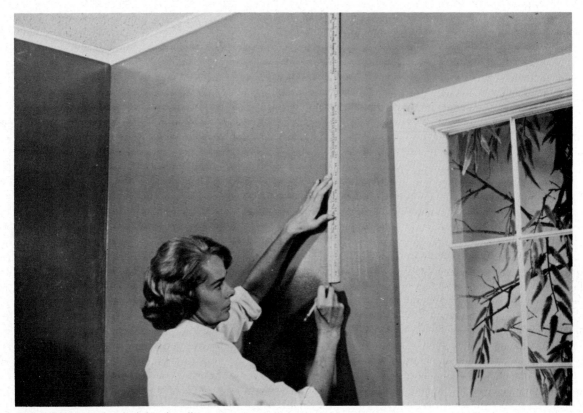

Fig. 8-2. Measuring height of wall.

the area of the walls. In practice there may be a baseboard and a molding, and the covering is to go between them, so this is the height you must measure. Use a yardstick, as shown in Figure 8-2. Assume the height is 8 feet. The area of the walls is thus 70′ × 8′, or 560 square feet. Divide by 30 to get the number of rolls. When you divide 560 by 30, you get between 18 and 19, so your first approximation is that you will need 19 rolls. However, you are not going to cover doors and windows, and you should subtract those areas. If you wish, you can measure everything exactly, but it is safe to say that an average window or door takes about half a roll. If your room has four windows and two doors, as in Figure 8-1, you can simply subtract 3 rolls from the 19 figured for the whole room. You should buy 16 rolls.

If the wallcovering is plain, you may use almost all of it without waste and cover almost 36 square feet with each roll. If your dealer will permit, arrange beforehand to return any unopened rolls for refund. Then you won't have to worry about whether you bought enough or too much for the job.

If you intend to cover the ceiling also, then you need to know the ceiling area. If the room is irregular, divide it into rectangles and add the areas of all the rectangles. The ceiling of Figure 8-1, as indicated by the dotted line, can be divided into one rectangle 14′ × 16′, or 224 square feet, and another 5′ × 6′, or 30 square feet. The ceiling area is then 254 square feet, and if you divide by 30, you see that it will take another nine rolls.

Borders are sold by the running yard, rather than by area. If you plan a border on all or part of the wall, measure the length of the part to be covered by the border, and simply give this dimension to your dealer.

A mural may be your choice for a large, unbroken wall. Measure that wall carefully, and take the dimensions to your dealer. The strips or panels of a mural must be matched exactly, and you do not have the latitude you

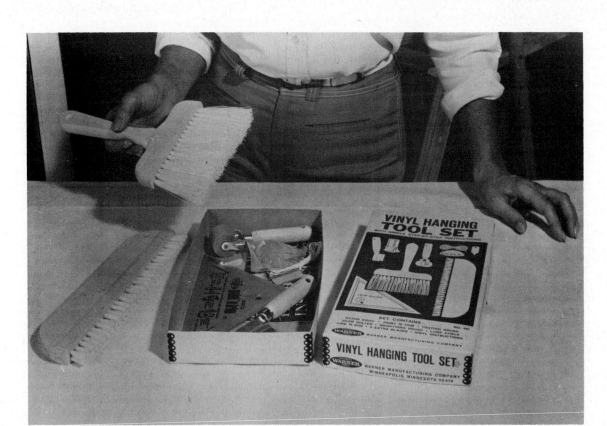

Fig. 8-3. Hanging kit.

have with a repetitive pattern. It is best to sketch the room with walls flat and plan exactly how the mural is to appear.

8-2. Tools

Tools for hanging wallcoverings are very simple and cheap. You will need a paste brush, a smoothing brush, a seam roller, a plumb line, chalk and a razor knife with extra blades. All of these are included in a hanging tool set, obtainable at your wallpaper dealer. A typical kit is shown in Figure 8-3. You will also need a paste bucket, a stepladder, a table, a screwdriver to remove electric fixtures, a ruler, scissors, a bowl and sponge and sandpaper. These are not included in the hanging kit since they are usually available around the house.

You may not need all the tools in the kit. For example, if you use a prepasted covering, you will not need a paste brush. However, the cost of the kit is less than it would cost you to buy each of the tools separately, so you might as well buy the kit. If you use prepasted wallcovering, you will not need a paste bucket, but a water tray, which is also inexpensive. Unlike painting tools, bargain hanging tools will work just as well as expensive ones.

8-3. Selvages and Seams

When wallpaper is printed, the design is not carried out to the edge of the bolt, but a small strip is left on each side. The strip is called a selvage, and it should be removed before the wallcovering is hung. Most modern coverings have selvages removed at the factory and are call *pretrimmed*. Some have perforations along the selvages. To remove the selvages, whack the tightly rolled bolt against the edge of a table, and the selvage will part at the perforations. Some delicate wallcoverings are neither pretrimmed nor perforated. The selvages are left on these for added protection.

When you hang paper, the simplest way to make a seam is to butt one strip against the next without any overlap. This is called a butt seam and requires trimmed edges. It is the preferred method of making seams since there are then no double thicknesses of covering on the wall. If you are worried about butting strips exactly and fear an opening may allow the wall to show through, you can overlap the strips slightly, about 1/16 inch or less. This is called a *wire-edge* seam. The double thickness at the seam is only slightly noticeable. An *overlap* seam is used when only one selvage is removed. Each strip must cover the untrimmed selvage of the preceding strip.

8-4. Preparing the Wall

Before hanging a new wallcovering, you must remove any old paper. At one time, new wallpaper was simply placed over old, and it was not uncommon for a wall to have three to five layers of wallpaper. The dry-strippable coverings, however, should be placed on bare walls and not over other papers. If the old paper is strippable, pry off one corner and pull it from the wall. It should come off easily. The procedure for removing old paper that is not strippable is described in the next section.

Remove all switch plates and outlet covers as shown in Figure 4-2. Later if you wish, you can cover the plates with a small piece of the same wall covering material so that the switch will not be conspicuous. To do this cut a piece somewhat larger that the plate, matching the pattern to that already on the wall, and paste it onto the plate with any kind of adhesive. Tuck under the excess and screw the plate back on the wall, after cutting a hole for the switch button. Alternatively, you can replace the plate with an artistic plate that is compatible with the design of the wallcovering.

Before hanging the wallcoverings, make sure the wall is in good condition. Repair gouges and cracks with ready-mixed patching compound applied with a putty knife, as shown in Figure 4-1. Sand down any rough spots. Finally, wipe or wash the walls to remove grease or dust.

8-5. How to Remove Old Wallpaper

If the old wallpaper is not dry-strippable, it must be moistened and scraped off. You can use a mixture of warm water and vinegar, or special wallpaper remover preparations that are mixed with warm water, and apply the mixture with a sponge or mop to loosen the old paper. However, the fastest method is to use a wallpaper steamer, which can be rented from most dealers. The rental cost is small compared to the cost of the wallcoverings.

A steaming machine has a boiler in which water is heated, by electricity in small units, or by kerosene in larger ones. The steam from the boiling water is fed through a hose to a perforated plate which is held against the old wallpaper. The steam soaks into the paper, softening the paper and the paste holding it to the wall. The paper can then be stripped off the wall with a putty knife or any flat tool. A special tool for the purpose, called a *wall scraper,* looks like a very wide putty knife.

When using the machine, begin at the bottom of a wall, since the hot steam tends to rise and will soften the paper above the plate as well as directly under it. Move the plate upward slowly with one hand while you strip the paper below with the other. Make sure the room is well ventilated to prevent it from filling up with steam.

The steamer can also be used to remove wallpaper from a papered ceiling. However, if paper is to be removed only from walls and you want to leave the old paper on a ceiling, you cannot use a steamer. The rising steam would loosen the ceiling paper. In this case you will have to soak the paper on the wall by hand. When the paper is soaked, you will be able to strip it off with a flat tool.

After removing the old paper, wipe up all excess moisture as soon as possible. Patch the wall where necessary and clean and smooth it before hanging the new wallcovering.

8-6. Cutting the Strips

Before you cut your bolt of wallcovering into strips for hanging, you should uncurl it so that it will lie flat. Unroll about three feet of the material and drag it firmly over the edge of a table to give it a reverse curl. Be careful not to crease the material. Repeat this step until the material lies flat.

Cut the first strip from the roll about four inches longer than the height of the wall. However, you cannot just cut similar lengths for the rest of the strips unless you have plain paper or a pattern that can be matched at random. You must be sure that the second strip is long enough so that if it must be raised or lowered to match the first strip, there is enough extra at top or bottom to cover the wall. If the pattern has a large vertical repeat so that there might have to be a large amount of waste material, you can sometimes cut down the waste by cutting alternate strips from different rolls. That is, cut the first, third, fifth strip, and so on from the first roll, and cut the even numbered strips from another. If you do this, you might want to number the strips on the back at the ceiling end so that you can keep track of the proper order of hanging them.

You should have a large table to work on, but if not, you can improvise by placing a wide plywood board 5 or 6 feet long on a bridge table. If you have a longer board, put it across two bridge tables for a firmer support. The width of the board or table should be at least double the width of a strip. Place the first strip on the table with pattern side up. Cut enough strips to do one wall, always making sure you leave a few inches extra at top and bottom and enough additional length to make sure the designs match from strip to strip. As each strip is cut, place it on top of the pile of strips

with the pattern side up and all heading in the same direction. You will have to cut some short strips to go over doors and over and under windows. These short strips must be placed in the pile in the proper order. Now turn the whole pile over so that the top strip is the first to be hung, and all are in the proper sequence. Each strip will now have the pattern side down so that it is in the proper position for pasting. Push the whole pile to the back of the table.

8-7. How to Apply Paste

Paste should be mixed from 15 to 30 minutes before it is to be applied. Mix the powder with water in a plastic paste bucket according to directions. Tie a string across the top of the bucket by punching holes near the top on opposite sides. You can then rest your brush on the string when you are not using it and thus keep the handle dry. When buying paste, make sure you buy the type recommended for the wallcovering you will use. After mixing the paste, spread newspapers on the floor near the walls and underneath and around your pasting table.

Pull the first strip off the pile to the front of the table still with pattern face down. Have the strip stretched over the table so that its upper end, the ceiling end, hangs over the edge of the table. The bottom end of the strip should be on the table near the opposite edge. Start brushing the paste on from the bottom of the strip. When the strip is pasted over half its surface, fold the bottom edge over, without creasing the paper, so that pasted surface is against pasted surface. This step, called *booking,* is shown in Figure 8-6.

Slide the wallcovering along the table so that you can now apply paste to the rest of the strip. Leave the top inch or two unpasted. If the strip is very long, you may want to book the upper portion, too, so that you can carry it without touching the floor. The strip is now ready for hanging. Hang each strip before applying paste to the next one.

Fig. 8-6. Booking.

8-8. Hanging the Strips

As you hang each strip, you must be careful to match the design to that on the strip just preceding it. When you have gone around the room, it would be a rare coincidence if the pattern on the last strip matches the first. Since you must expect a discontinuity in the pattern, try to have it where it will not be noticed. Thus, you would normally begin at the edge of a door and proceed around the room from there. Anyone entering through that door would have the discontinuity behind him and would be less likely to notice it. If there is a large window in the room, you can begin at the edge of it, and let the window itself be the discontinuity in the pattern. Of course, if you are not hanging wallcovering on all walls, you do not have to worry about matching where first and last strips meet.

The first strip must be hung exactly vertically. Do not depend on the walls or edge of a door to be a true vertical, since houses are not built that exactly, and even if they were built to true verticals, there is always some sag and displacement. To locate the first strip you must mark a true vertical on the wall. Do this before you apply paste to the first strip. Measure the width of the strip. Now measure out from the door or window where the first strip is to go a distance one inch less than the width of the strip. Drive a tack in the wall near the ceiling at this distance from the door. Attach your plumb line to the tack, and rub the line with chalk. When the plumb line bob comes to rest, grasp it firmly without moving it, and with the other hand snap the line against the wall, as shown in Figure 8-7. You will now have a true vertical line on the wall. Every time you have to go around a corner, repeat the operation to get a new vertical on each wall.

Grasp the first strip near the top and climb your stepladder. If the top was booked, unfold it. Hang the paper so that the top overlaps the

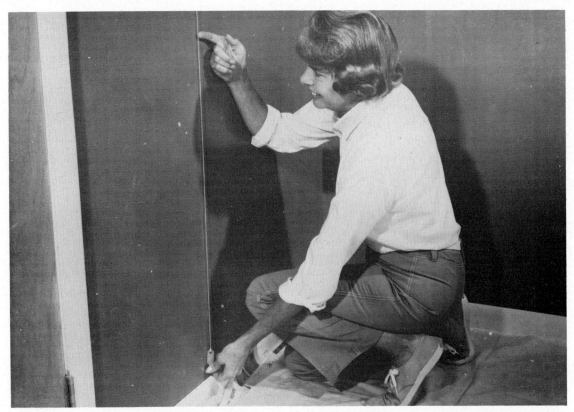

Fig. 8-7. Vertical line.

Fig. 8-8. Hanging the strip.

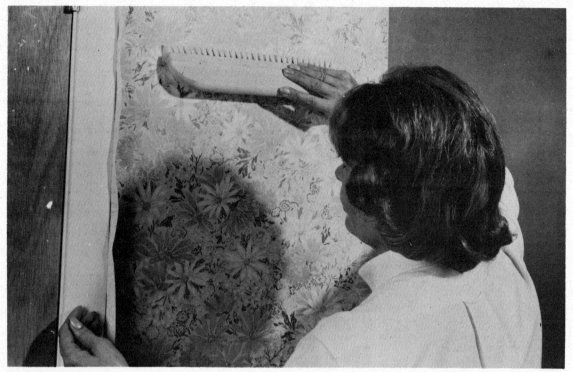

Fig. 8-9. Smoothing.

Fig. 8-10. Trimming excess.

Fig. 8-11. Sponging off paste.

ceiling joint about two inches, as shown in Figure 8-8, and so that one edge runs right along the vertical line. Use a smoothing brush, not your hand, to rub the paper against the wall, as shown in Figure 8-9. Use downward strokes from the center of the strip to the edges. When the top of the strip is smooth, unfold the bottom of the strip and smooth the rest of the strip to the wall.

After the bubbles have been smoothed out of the strip, cut off the excess at door edges, baseboards and ceiling, while the paper is still wet. The simplest way to do this is to use a wallscraper or putty knife as a guide and cut the material with a sharp razor edge, as shown in Figure 8-10. Before hanging the next strip, sponge off paste from baseboards, ceiling and from the wallcovering, using clear water, as shown in Figure 8-11. If the paste dries, it is difficult to remove.

Hang each succeeding strip in the same manner, using the edge of the preceding strip as a guide. To move a strip, as for matching or lining up, place your palms lightly near the center and push up, down, or sideways as required. Do not grasp the strip at the edges to move it or the paste will get on your hands.

Hanging paper behind a radiator may be a problem because you cannot get at the material with your smoothing brush. Use a long stick, such as a yardstick, with cloth wrapped around it. The same technique is used in any tight place.

Hang your wallcovering right over electrical outlets and switches after first removing the covers. When the paste is dry, cut away the paper around the opening. If there are wall brackets, they should be disconnected and removed before hanging the paper. If this cannot be done, cut a slit in the material from the edge to the location of the wall bracket, and slide each side of the slit under the bracket. Some additional trimming may be necessary, but you can paste all edges down so that the slit won't show.

Corners present some difficulty because they are not always square. When a strip is on one wall and part of the next, figure out

Fig. 8-12. Trimming around window.

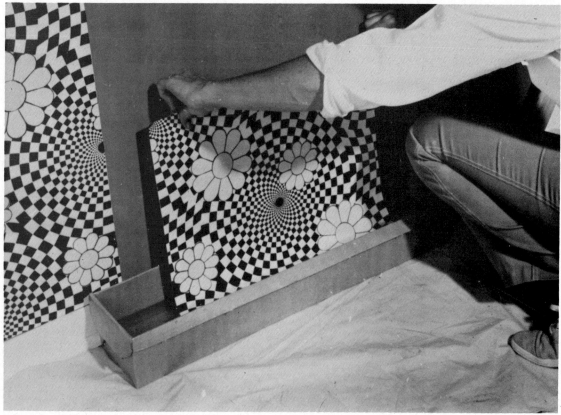

Fig. 8-13. Prepasted wallpaper.

approximately where the edge of the strip will be. Draw a vertical chalk line as a guide at that point. Now hang the strip, smoothing it carefully on the first wall and for about one inch of the second. Make sure it is snug in the corner. If the outside edge is not true with the chalk line cut the strip about one inch from the corner and start a new strip so that the edge is lined up with the vertical. There may be some overlap or mismatch at the corner, but it will not be noticeable. Each time you go around a corner, always set a new vertical reference line.

Hanging wallcoverings around windows or doors is not much different to the procedure on other walls. Line up the strip to match the preceding strip. With scissors, make a diagonal snip in the strip at the corner of the frame. Fit the wallcovering in place and trim off the excess with a razor, using a scraper as a guide, as shown in Figure 8-12.

After strips have been in place for about 15 minutes, roll the seams with the roller in your tool kit. In practice, you might roll the seams after hanging three or four strips. You must not roll the seams of flocked or embossed coverings, however, because you might crush them. For these papers, rub the seams lightly with a damp sponge.

8-9. Hanging Prepasted Wallcovering

If you are using prepasted coverings, you do not have to mix and apply paste and you do not need a table. However, you will require a water tray. Otherwise, the procedure is similar to that for pasting and hanging.

Cut the prepasted strips in the manner described in Section 8-6. Reroll each strip loosely from bottom to top with the pattern inside. The strip is placed in the water tray next to the baseboard at the place where it is to be hung, as shown in Figure 8-13. The water should be tepid. Follow the manufacturer's instructions about water temperature and length of time for soaking. Pull up the wallcovering from the water tray, as shown in the figure, and hang as described in Section 8-8.

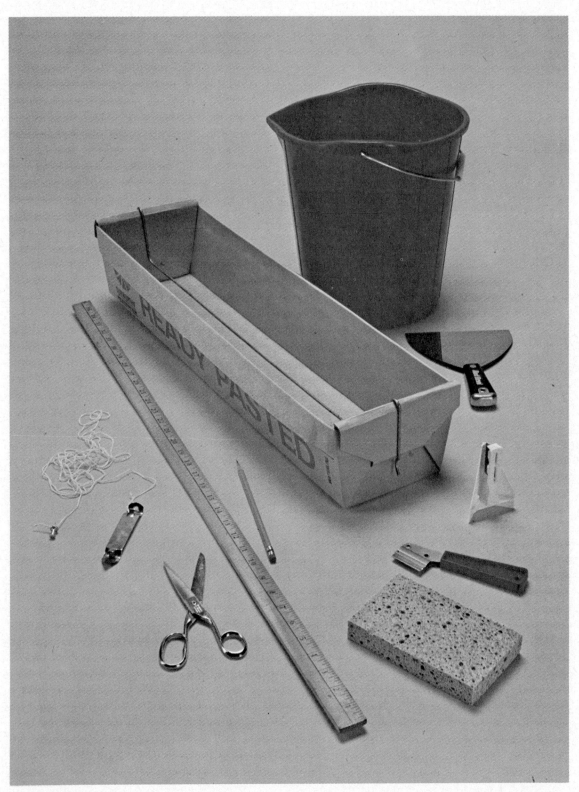

Paper-hanging tools.

Excess paste may be sponged off edges.

Rigid Wallcoverings: Panels

Not too long ago, panelled walls were synonymous with luxury. The middle classes and poorer folk had to do with painted walls or wallpaper, but the very rich had wood on their walls. Panels ranged from formal, luxurious mahogany to informal, but still expensive, knotty pine. Each board was hand cut and fitted in place, contributing to the expense. Today, new materials and simplified installation techniques have reduced the cost of panelled walls so that this former luxury is now within everyone's means.

9-1. Materials

Instead of hand-fitting individual boards, the handyman works with large panels that cover big areas and are easily installed. Most panels are 4' wide and extend from floor to ceiling. A common size is 4' X 8', but lengths of 7', 9', 10', 11' and 12' are also available. You can generally find a size that fits your walls, so that sawing and trimming are minimized.

Hardboard panels are made of selected wood chips pressed together with a suitable binder to form a hard panel of the proper size. The hardboard is then coated with a synthetic finish that resembles wood grains, tile, leather, cork, marble, or almost any other material, as well as murals and a variety of original designs. The wood designs are realistic, representing vertical boards butted against one another. The "widths" of the boards, that is, the spacing between vertical grooves in the design, are random, as would be the case if individual boards were used. This also has the advantage that when one panel butts against another, the line of demarcation looks like just another groove. Some designs of antique wood faithfully reproduce the wormholes in the wood, as well as the "defects" that one might expect to find in weathered wood.

Textured designs in hardboard panels look and feel like the authentic material. Simulated cracks in marble or old leather are there to touch as well as to look at. Other designs, such as abstracts, may be smooth or rough, according to the whim of the designer. Before choosing a material, look at samples of different panels from different manufacturers so that you cover a wide selection of types and styles.

Wood-veneered plywood panels start with plywood sheets of the proper size. The plywood is covered with a thin veneer of desired wood, and the finish not only resembles natural wood, it is the actual wood. In some of these panels, the wood veneer is impregnated with a transparent plastic so that it never needs maintenance. Even without the plastic, the wood veneer requires only an occasional dusting to maintain its finish. The veneer is not an imitation. It is a thin layer of natural wood, showing grain, knotholes, and other "blemishes" in the original material. These blemishes are included to add interest, as accents, in what might be an otherwise monotonous expanse. This explains the attractiveness of knotty pine.

Wood-veneered panels come in natural wood colors and may also be stained with any of the usual wood stains. Hardboard panels with synthetic finishes can be manufactured in any color. Some are made in true natural colors, while others are in decorator colors.

When installing panels, especially if walls are not exactly vertical, you may find a gap at the top or bottom of a panel or at the junction of two panels or the junction of a panel and a door frame. To cover these gaps, and to dress up a panelling job in general, manufacturers offer an assortment of moldings in colors to match the panels. The molding may be made of the same material as the panel or may be wood or aluminum covered with a veneer of matching material. Cross sections of some of the shapes are shown in Figure 9-1. Others in this line include inside corners, spacers, or dividers, chair rails, sills, and almost any other shape you might have use for in covering or hiding an imperfection in the work. Each molding comes in lengths as long as the panels and is cut to size as needed for shorter runs. Some typical installations are shown in Figure 9-2. In (a), a base molding is used to cover the gap between the bottom of a panel and the floor. A shoe molding is used to dress up the joint. In (b), a cove molding is used to cover the same sort of gap at ceiling level. When a panel is used to cover only part of a wall, the top edge of the panel looks

unfinished. For the sake of appearances a cap molding may be used to cover the edge, as in Figure 9-2 (c). When you are selecting panels, look over the wide selection of moldings available.

Panels and moldings can be attached either with special adhesives or with finishing nails. Both are available from the manufacturer. Nails have heads that match the finish of the panel so that there is no need to countersink them. In addition there are putty sticks in colors to match the finishes of the panels. These may be used to touch up scratches and minor imperfections. The stick is simply rubbed over the scratch, and the spot is wiped with a dry cloth. The material hardens quickly and hides the blemish completely.

9-2. Deciding to Panel

There are many good reasons for considering panelling when a wall needs repair or refinishing, but your first reason must be that you like panels. If you don't like the look or feel of wood on the walls, there is no sense considering the advantages of panelling over other types of finish. Indeed, there is a place for panelling in every home, but also a need for other wall decorations.

If you are building a new room such as a basement playroom, or are converting a garage to a family room, panelling will save some work. Walls can be made of panelling fastened to the studs and will need no other finish. Another place where panelling may be best used is in any room with severely damaged walls. Panelling can be installed

(a) Base   (b) Shoe   (c) Cove   (d) Cap   (e) Outside corner

Fig. 9-1. Cross section of moldings.

Tools for installing panelling.

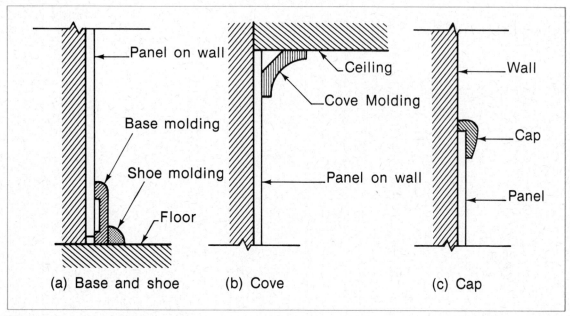

Fig. 9-2. Molding installations.

right over cracks, holes, or other defects.

Panelling is relatively easy to apply. Complete installation instructions are presented in Chapter 10, and after reading them, you should be able to tackle the job with confidence that the result will be quite presentable. There is no messy clean-up after the job is finished. If you saw panels indoors you will have to pick up the small amount of sawdust with your vacuum cleaner and put away your tools, but that is all the cleaning necessary.

Panelling is not more expensive than other wallcoverings. Inexpensive hardboard panels cost about the same as wallpaper. Wood-veneered panels are more expensive, but not more than high-quality wallcoverings of other types. Price in panels does not depend on type of wood or color, but rather on the method of fabrication. Many of the wood grains available in wood-veneer are also available in hardboard, and the hardboard will pass all but the closest scrutiny.

As mentioned in the preceding section, panels are available in a variety of colors. Thus, panelling a room does not mean that it will necessarily be dark, although original wood panels were only of dark wood. You can have any color you need to blend with your furnishings. However, you would not want a solid unbroken stretch of four panelled walls in a room any more than you would want plain, painted walls. You should decide to add accents perhaps by a combination of panelling and a patterned wallcovering.

One advantage of panelling is ease of maintenance. Plastic-impregnated panels can withstand abuse and are easily cleaned of dirt, grease, and even wax crayons and ink. The finish lasts for years, literally for as long as the wall lasts.

Panelling can be done piecemeal. You can panel one wall or even part of a wall, leaving the rest in wallpaper or paint. There's nothing wrong with a wood panel accenting a painted or patterned wall. Later you can add additional panels as you see fit, to make a solid panelled room, or to leave a smaller amount of wallpaper showing so that the design becomes an accent in the panelled room.

9-3. How to Use Panelling

The general principles of color harmony and design, explained in Chapter 7, apply equally designers are reproduced, providing a rich variety of imaginative designs for your walls. In the case of panels the concepts of professionals are also reproduced, although the variety is not as great. However, with imagination there is no limit to the ways you can use both panelling and flexible wallcoverings. Indeed, the type of panelling or design on the wallcovering is just a starting point from which you can create the room of your own choosing. Be imaginative and inventive; don't be bound by conventional uses of the materials.

As indicated in Chapter 7, a predominant color on the wall accentuates furnishings of the same color in the room. Going one step further, a material that predominates on the wall accentuates furnishings of similar materials in the room, as shown in Figure 9-3; a knotty red cedar wood-veneered panel adds a touch of simple elegance to a wall and brings out the best qualities of the fine wooden furniture.

The treatment in Figure 9-4 is somewhat different in that panelling is used on only one wall. Each large wood-veneered panel is "sculpted" so that it appears as an accent adding to the quiet luxury of the room without monopolizing the décor. The result is a formal, but still warm and interesting, conversation area.

An unconventional arrangement of panels is shown in Figure 9-5. Two different hardboard panels are used on the walls of this studio apartment. A rich-looking rosewood panel is used in normal fashion in part of the apartment, and a white panel is mounted horizontally on another wall. In addition, painted hardboard is used on the ceiling to provide accents. Note how the horizontal panelling creates the illusion of a room separate from the rest of the apartment. For striking effects, panels can be used diagonally as well as horizontally. Lines on adjacent panels need not be in the same direction.

Panelling is most efficient when a new room is to be created. Two views of a recreation room built in a former garage are shown in Figures 9-6 and 9-7. The hardboard shelving and panelling in Figure 9-6 have simulated brown oak finishes which accentuate the color of the hexagonal coffee table. If the wall

Fig. 9-3. Wood on wood.

to panelling. Dark colors make walls seem closer and smaller, while light colors make a room larger. Vertical lines on a wall make the ceiling seem higher. The apparent shape of a room can be altered by choice of color and design.

Imagination is an important ingredient in interior decorating. In the case of flexible wallcoverings, the conceptions of professional

Fig. 9-4. Living room wall.

Proper choice of color for the walls of a room can enhance the beauty of fine furniture.

The colors of the appliances, the counter and the walls are nicely coordinated in this kitchen.

Fig. 9-5. Studio apartment.

Fig. 9-6. Recreation room built in garage.

A finished basement with panelling and stone fireplace.

Fig. 9-7. Oak wall with painting.

Fig. 9-8. Laundry.

were bare, the wide expanse of oak panelling might be too monotonous. The painting on the wall in Figure 9-7 breaks up the large area, but it also ties the room together by featuring colors found in the rugs, drapes and chairs.

Another newly created room is the laundry shown in Figure 9-8. Here the wood-veneered panel was chosen for durability and imperviousness to water vapor. This simple panelling will probably outlast the laundry equipment. Note that in a work area such as this, you are usually more concerned with the practical attributes of the wallcovering rather than with its decorative value. No special care is taken to make the wall more interesting, although the contents of the horizontal shelf do break up the large uniform area of the wall.

The bathroom walls in Figure 9-9 are also covered with wood-veneered panels that withstand water vapor. In contrast to the laundry and other work areas, lavatories should have interesting walls. The mirror, painting and towel ring, and the contrasting patterns of adjacent walls furnish the needed accents.

Bathrooms conventionally have walls covered with tiles or marbled plastic. Hardboard panels covered with tough plastic are available to simulate tiles, solid colors, and other common bathroom walls. The finish is heat resistant and impervious to moisture so that these panels can be used even around the tub and shower. A typical installation is shown in Figure 9-10.

Dining rooms should be interesting but not exciting. A simple, but formal treatment is shown in Figure 9-11, where rough textured white hardboard panelling covers most of the wall area. Geometric patterns add interest, but are not too stimulating for formal dining.

The pitfall with a child's bedroom is that he or she might outgrow the decoration. Since panelling lasts for years, it is important to choose a design that will not be too childish for a teenager or a young adult. In the boy's room shown in Figure 9-12, the wall is a rugged white hardboard built to stand abuse. As the boy grows, new decorations reflecting his new interests will replace the old, but the wall is suitable for any age group. The storage unit is also built of hardboard panels.

The girl's bedroom shown in Figure 9-13 illustrates an imaginative use of panelling beyond simple wall covering. The built-in canopy bed utilizes the same wood-veneered panels as the walls. Note that the panels that frame the canopy match the curvature of the panels that frame the bed. The pieces were cut from the same panel without waste. In this room, as in the boy's bedroom in Figure 9-12, the walls will last throughout the girl's childhood and will still be suitable for her as a teenager.

An attic room with a sloping ceiling can present a decorating problem, but panelling can eliminate a major portion of the difficulty, as shown in Figure 9-14. The large attic room shown has a sloping ceiling at each end. The portion of the room under the sloping ceiling at one end was made into a walk-in closet with doors made of the same woodgrain hardboard that was used on the walls of the room adjacent to that end. The textured white hardboard shown on the ceiling was carried over to the far wall (not shown) so that the room was effectively divided into two distinct areas.

Recreation rooms or playrooms are frequently panelled because the room is an addition, perhaps in a basement, and panelling is the most efficient method of combining construction and decorating. However, by proper choice of materials, a recreation room can be as formal or informal as you would like to make it. The formal recreation room in Figure 9-15 is designed as a showcase for the hobbies and interests of the occupants. White panelling on the walls creates the illusion of greater space and does not compete for attention with the collections. Rosewood hardwood provides a rich contrast, drawing attention to the individual collections separately.

An example of what can be done with a basement recreation room is shown in Figure 9-16. A light-colored hardboard panel simulating antique wood was used on three walls and around the bar. Important considerations in choice of materials for recreation rooms are durability and low maintenance. One wall was covered wth a bright plaid wallpaper to set apart a conversation area. Flooring is a brick-

Fig. 9-9. Bathroom.

Fig. 9-10. Marbleized bathroom.

Fig. 9-11. Dining room.

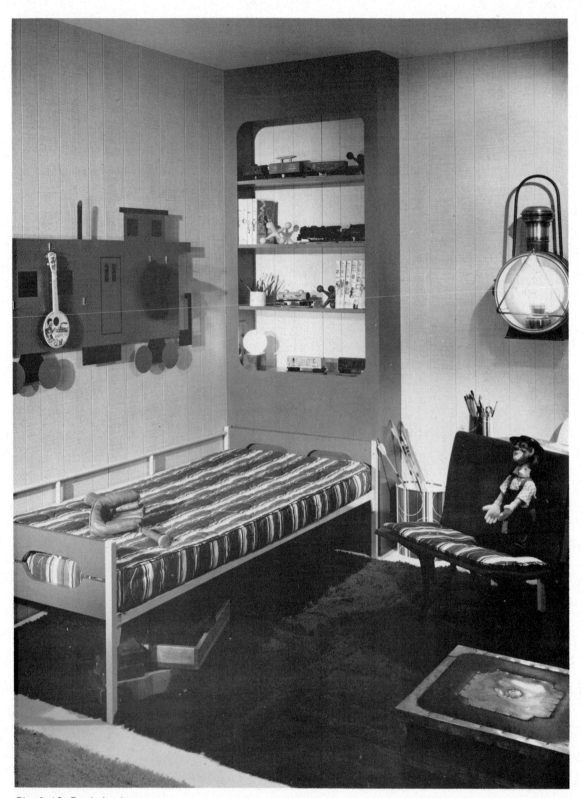

Fig. 9-12. Boy's bedroom.

Fig. 9-13. Girl's bedroom.

Fig. 9-14. Attic room with sloping ceiling.

Fig. 9-15. Hobby room.

Fig. 9-16. Basement recreation room.

Fig. 9-17. Recreation room.

patterned no-wax tile, again for easy maintenance. To soften noise, a shag carpet is used on the floor and a white acoustic tile on the ceiling.

When installing panels, you will have a certain amount of sawing to do, to fit pieces around doors and windows or to cut openings for ventilators and electric fixtures. Save all the odd scraps and use them for decorating other parts of the room, as shown in Figure 9-17. The active areas in this recreation room are covered with a vinyl-covered wood-veneered panelling that withstands abuse and needs no maintenance. One wall is painted a solid color and is accented with odd pieces of the same panelling.

10 How to Install Panels

The biggest problem in installing panelling is lack of confidence. Don't worry! The manufacturers have built the expertise into the product, including materials to correct errors. Read the instructions carefully before you begin. You will see that the installation is not difficult.

10-1. Ordering and Storing Panels

When you buy panelling, you probably won't be able to arrange to return any unused panels for credit. Consequently, you must be able to make an accurate assessment of how many panels you will need. The "standard" panel size is 4' × 8', and the "standard" wall is 8'. Thus, a standard panel will reach floor to ceiling in an 8' room. If your ceiling is higher or lower, you can buy panels that are longer or shorter, but the cost per square foot for the 8' panels is lower than for the others. If you are panelling less than the full height of the wall, you use the height of the desired panelled section as the height of your room.

If you are panelling all the walls of a room, first measure the perimeter. This is the sum of the width of all four walls. If you are panelling only three walls, or parts of walls, then measure the width of the walls or sections you plan to cover. Divide the perimeter (in feet) or the width you will cover by the number of walls you plan to panel and this will give you the number of panels that would cover the walls without regard to windows, doors, fireplaces, and other openings that would not be covered. Subtract 1/2 panel for each door or fireplace, and about 1/4 panel for each window. These are approximations, and if you wish, you can measure openings accurately. However, if you want to panel above and below windows and above doors, you should figure how these odd pieces can be cut from one panel. If you use approximations for openings you will probably arrive at the correct figure for the number of panels, but if your doors and windows are odd sizes, you should plan where each panel or piece of panel is to go and thus get an accurate figure for the number of panels needed.

When the panels arrive, store them in a dry place before installation. They should be stored flat to prevent warping. Panels may be shipped in large bags and packed in pairs with the faces of each pair toward each other and separated by a protective sheet. When removing the panels from the bags, slide them out two at a time, keeping the protective sheet between the adjacent faces, so that the finish on each panel is protected. Separate the panels and stand them in the room where they will be installed. Leave them there for at least two days to become acclimated to the temperature and humidity in the room.

When you stand the panels in the room, you may notice differences in color, grain, or pattern of knotholes. You can arrange the panels around the room to get the most pleasing effect. When you have arrived at a suitable arrangement, number the panels on the back in the sequence that you plan to install them on the walls so that you can keep track of them more easily.

10-2. Tools Needed

No special tools are needed to install panels, but you should have a few common tools handy so that you don't have to look for them when you need them. You may not need all of the following, but have them available:

- Hammer

- Nailset

- Tape measure

- Saw (hand or power)

- Level

- Small wooden block

- Scraps of shingles or other wedge-shaped scraps

- Caulking gun for adhesive (this is supplied with the adhesive)

- Small compass for scribing

10-3. Fastening Methods

Panels can be fastened to the walls by nails or adhesives, and each method has its advocates who claim it is superior to the other.

Either method gives satisfactory results, but the two techniques are somewhat different.

When using nails, it is necessary to locate the studs in the wall. The nails are driven into the studs to give sturdy support for the panels. Panels can be nailed directly over old wallpaper, on greasy or dirty walls, on walls with gaping holes in them, or to studs in new construction. One disadvantage is that each nail must be driven in separately with a hammer and nailset, and this may seem tedious.

When using adhesives, it is not necessary to know where the studs are, but the wall must be in suitable condition. You cannot use adhesive over wallpaper, since the adhesive may loosen the wallpaper paste and pull the paper off the wall. Walls must be cleaned, since the adhesive will not stick to grease or loose dirt. Adhesive can also be used to stick panels to studs in new construction.

The supporters of adhesives claim it is faster and simpler than nailing. However, the total time for the job is more a matter of experience than method of fastening. There is no doubt that nailing can be used in some situations where adhesives may prove unsatisfactory, but with proper care, both methods provide strong bonds to the wall.

10-4. Installation with Nails

You can install panels from floor to ceiling or you can butt them against existing molding. If you choose to panel the whole wall, you must remove the old molding. Just pry it loose and don't worry about damaging the walls, since the panelling will cover any holes you make. In either case, measure the space where the panel is to go as accurately as possible. Due to the settling of the house, the height from floor to ceiling may vary around the room. Small differences are not important, since gaps can be covered by moldings at floor and ceiling. The main reason for measuring is to trim the panels for installation where the ceiling is lower. The panel should be 1/2″ to

Fig. 10-1. Drawing line over stud.

1-1/2″ less than the overall height to allow for expansion caused by moisture and temperature variations. Moldings will cover the gaps.

The panels will be nailed to the studs. In new construction the studs are visible, and the panels are fastened directly to them. When remodelling, it is necessary to locate studs in the wall. You can tap lightly on the wall and should hear a hollow sound between studs and a solid sound when directly over a stud. You can also use an inexpensive stud locater. The studs should be spaced 16″ apart between centers, and when you have found one, you simply measure at 16″ intervals to locate the others. Drive a long nail lightly into the wall where you think a stud should be. There should be resistance but if it goes in too easily it is between studs. Again, don't be concerned about nail holes or other defects, since the new panels will cover them. Probe with the long nail on both sides of the stud until you have located it exactly. Draw a vertical line over each stud. The simplest way to do this is with a chalk line, as shown in Figure 10-1, but you can use a yardstick and level.

Start in a corner and butt the first panel against the adjacent wall. The outer edge should fall on a stud and be parallel to the vertical lines. If the outer edge is not parallel, the intersection of the two walls is not exactly vertical. Shim the panel up from the floor by using wedges or scraps of shingles so that there is a gap of about 1/2″ at top and bottom. Make sure the outer edge is vertical, either by using a level or by checking it with the vertical lines you drew. If the gap at the wall is small at its widest part, you will be able to cover it with corner molding. However, if the discrepancy is more than about 3/4″ you will have to trim the panel. Use the compass to scribe the amount to cut off. With the outside edge of the panel vertical, and the panel touching the corner at one point, place the compass point against the adjacent wall and, holding it in contact with the wall, draw a line on the panel from top to bottom, as shown in Figure 10-2. In addition to correcting for an irregularity at the corner, it may be necessary to trim the panel so that the outer edge falls on a stud.

The panel can be cut with a handsaw or power saw. With a crosscut handsaw or table saw, place the panel face up, but with a sabre saw or portable power saw, it should be face down to avoid splitting the veneer. Do not use a ripsaw. Both sides of the cut should be supported so that the panel will not split near the end of the cut. You can lay the panel on four sawhorses or on two bridge tables. Sawing a panel is illustrated in Figure 10-3. When it is necessary to cut an opening in a panel, as for an electric outlet, drill pilot holes first at the corners of the opening, then cut with a sabre saw or keyhole saw.

When the first panel is in its proper location, you should nail it in place. Start at the corners, as shown in Figure 10-4. All nails should go into studs. For new construction with panels mounted directly to studs, use nails 1″ to 1 1/4″ long. The same lengths are used when nailing to furring. (See Section 10-6.) For covering old walls, use nails 1-5/8″ to 2″ long. The edges of the panel should lie along studs, and nails should also be driven into the studs that lie between the edges. You can use ordinary finishing nails and countersink them. Then fill the holes with colored putty to match the finish on the panels. This putty comes in a stick which you can buy when you get the panels. You can also use nails with colored heads which don't have to be countersunk because they match the panels.

When panels are fastened directly to studs, nails should be spaced about 6″ apart along the edges of the panels and about 12″ apart elsewhere. Over furring, these spacings can be increased to 8″ and 16″ respectively. Over old walls, space the nails every 4″ at the edges and about 6″ elsewhere.

After the first panel is nailed in place, butt the next panel against it, using a wooden block at the edge to tap it snugly against the first. As with the first, use wedges or pieces of shingle to keep the second panel about 1/2″ off the floor. Nail this panel in place as you did the first, and continue with the rest of the panels. Note that if you are panelling from an old molding instead of from the floor, the panels are butted right against the molding and are not shimmed.

After a few panels are up, you may come to

Fig. 10-2. Scribing at corner.

Fig. 10-3. Sawing panel.

Fig. 10-4. Nailing.

Fig. 10-5. Door cut-out.

a window or doorway less than four feet from the edge of the last panel. If, for example, only 3′ of space remain, simply cut 1′ off the width of the panel. Your saw doesn't even have to be straight since you can cover the joint with a molding. For the small space over the window or door, cut the proper filler from scrap pieces if possible. For example, after you trim the panel to fit into the 3′ width, you will have a piece about 1′ by 8′. Cut this into heights to fit over the door. For a fancier job, you can cut the panel to fit around the door, as shown in Figure 10-5.

When you come to the opposite corner, you may be lucky and find that the space from the edge of the last panel to the corner is exactly four feet and about 1/2″. (You should leave about 1/4″ to 1/2″ space to take care of possible expansion of the panels.) More likely you will have to cut a panel to fit, just as you did for the door mentioned in the preceding paragraph. Cut the piece and nail it, in the same way that you nail a whole panel. Now when you start on the next wall, begin flush with the corner. The first panel on the next wall will hide any irregularities in your saw cut. If you wish, you can cover the corner with molding to hide the joint completely, as shown in Figure 10-6.

10-5. Installation with Adhesives

Adhesives can be used to fasten panels on new studs, on furring, or on painted walls. The wall should be clean. Adhesives should not be used over wallpaper or on walls that are crumbling. The position of the studs with respect to the panels is immaterial, so they do not have to be located, as for nailing. The rest of the installation is similar.

Panel adhesives come with a caulking gun. Place the tube of adhesive in the gun and cut off the tip of the spout on the tube. With a long nail, puncture the seal at the bottom of the spout, and the gun is ready to use. To operate, pull the trigger holding the spout next to the wall and squeeze the adhesive out like

toothpaste. One tube could last for five panels if used sparingly, but it is better to use it generously and expect to put up only three panels with each tube of adhesive.

Fit the first panel into place in a corner, making sure the outer edge is vertical. Use a level or plumb bob to make sure the edge is correctly aligned. If necessary, trim the inner edge to fit in the corner, using the compass to determine the proper amount of trim, as shown in Figure 10-2. Have someone else hold the panel in place, or if no one else is available, wedge the panel in place with scraps of shingles. Then draw a vertical line on the wall next to the outer edge of the panel. Take down the panel and apply adhesive in a continuous line 1/2″ inside the perimeter of the whole panel, all the way around. Now add horizontal dabs of adhesive on the area inside the perimeter. These dabs should be about 3″ long and spaced about 6″ apart horizontally and 16″ vertically. Now press the panel in place. Make sure the outside edge is vertical and allow at least 1/4″ at top and bottom for expansion. (If you have left in the old base molding, you can butt the bottom of the panel against it.)

Pound the panel against the wall to squash the adhesive. You can use your fist, or a hammer on a cloth-covered block. Now pull the upper part of the panel from the wall, leaving the bottom edge touching, and rest the panel against the back of a chair. Allow the adhesive to set for about ten minutes and then push the panel back against the wall, again making sure the outside edge is vertical. Pound the panel again and press it firmly in place. Now you're ready for the next panel.

The rest of the installation is the same as for nailing. Doorways and corners are treated as described in the preceding section. Moldings are nailed in place after the job is finished.

10-6. Furring

If your old wall has bulges in it so that the panelling will not lie flat, or if you wish to panel over masonry walls, you will have to use

Fig. 10-6. Corner molding.

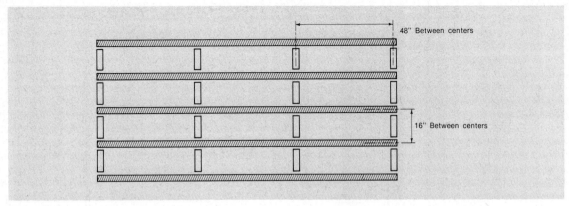

Fig. 10-7. Furring strips.

furring strips. These are simply pieces of 1″ × 2″ lumber nailed to the wall to provide a flat surface for supporting the panels. Horizontal strips are fastened along the whole width of the wall with a 16″ vertical spacing between centers. Then short vertical strips are added with a horizontal spacing of 48″ between centers. The arrangement is shown in Figure 10-7.

On plaster walls, the horizontal furring strips cross the studs, and the shorter vertical strips lie right on the studs. All the strips are nailed to the studs. The nailheads should be below the surface so that they won't interfere with the panelling. On masonry walls, use masonry nails to fasten the strips. If the original wall is uneven, use shims under the furring so that the surface of the strip is vertical. You can check this with a level.

After the strips are in place, you can install the panels either by nailing or using adhesives. If you use nails, they should be placed in all horizontal strips. Adhesive is applied to all the strips with the same spacing as for a plain wall.

10-7. Electricity Boxes

If an electricity outlet or switch or fixture lies on the area to be panelled, you will have to cut a hole in the panel to accommodate it. For outlets and switches, simply locate the posi-tion on the panel, and draw an outline of the box. Drill pilot holes at all corners and cut out the material with a keyhole saw or sabre saw. Your saw cuts need not be perfect because the wall plate overlaps the edges. However, you will have to pull the outlet from the level of the old wall to the level of the new panel. This requires longer screws in the junction box.

For a light fixture, you will have to disconnect the wires. Cut a round hole in the proper place in the panel, using a sabre saw or keyhole saw, with diameter smaller than the diameter of the fixture. After the panel is in place, connect the wires again and put back the fixture. Whenever you work near electric wires make sure the power is disconnected.

10-8. Care and Maintenance

Panels can usually be cleaned with a damp cloth. If necessary, use a mild soap or detergent to remove crayon marks and other stains. Do not use abrasive cleaners, as these may mar the finish. The surface may also be lightly waxed.

Although panels are unusually tough and wear-resistant, they can be abused. Light scratches can be removed by rubbing on a clear wax with the grain. Deeper scratches can be touched up with putty that matches the finish.

11 Masonry Panels

Brick and stone lend elegance to a room and they are materials which require no maintenance. Masonry panels are now available for householders to install their own brick or stone with very little effort and no prior experience.

11-1. Materials and Designs

Masonry panels are made of crushed limestone reinforced with fiberglass. The stone gives the panels an authentic texture, while the fiberglass provides strength and durability. A closeup view of a wall covered with man-made masonry panelling is shown in Figure 11-1. The panels have the appearance of real bricks but are much cheaper and have an additional advantage of being hollow so that the dead air space provides good thermal insulation.

Masonry panels weigh about one pound per square foot, which is heavier than wooden panels, but much lighter than the equivalent area of real brick or stone. One manufacturer makes three different types: the first reproduces weathered bricks, arranged twelve to a panel, as shown in Figure 11-2. The staggered bricks permit interlocking adjacent panels during installation. The second type

Fig. 11-1. Close-up of masonry panelling.

has twenty-four bricks to a panel and looks and feels like new sculptured bricks. The arrangement is shown in Figure 11-3. Both of these series are available in white, buff, and light or dark red. The third type reproduces the beauty of natural stone in a panel 1′ × 4′ in area. The panel looks like a section of wall made of stones of random sizes, as shown in Figure 11-4. It is available in white, grey and buff.

Masonry panels can be used in any room. You wouldn't want them for all four walls, but a suitable combination of brick wall or section

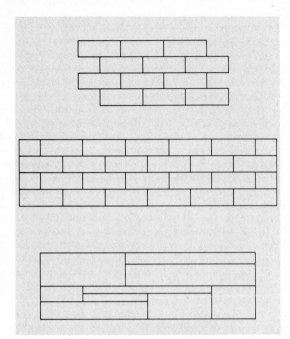

Fig. 11-2. "Heritage" panel by Masonite.
Fig. 11-3. "Carriage" panel by Masonite.
Fig. 11-4. "Bedford" panel by Masonite.

Fig. 11-5. Entryway with brick wall.

Fig. 11-6. Natural stone panelling.

with panelled or papered walls in the rest of the room can be an unusual and elegant touch. The brick panelled wall in the entrance in Figure 11-5 is inviting. A natural place to use masonry panels is in a playroom, as shown in Figure 11-6. The bathroom in Figure 11-7 features a red brick sitting area and an ornate black grill doorway leading to the dressing room. The brick is made of masonry panels, and the grill is in a decorator panel. The bath area is framed in stucco, which is made of hardboard panels with a stucco surface. Masonry panels can also be used around a fireplace, behind a counter in a kitchen, or on a wall of a dining room. They blend well with either formal or informal decorations.

The three types of panels shown in Figures 11-2, 11-3, and 11-4 are all designed so that adjacent panels interlock for ease of installation. The staggered end bricks in Figure 11-2 provide an obvious and simple method of interlocking panels horizontally. The panels of Figure 11-3 interlock vertically by means of a flange arrangement shown in Figure 11-8. The upper panel has a groove in its bottom edge, into which the tongue of the lower panel fits. Note that the fastener for the lower panel is then completely hidden by the upper panel. The stone panels of Figure 11-4 have an overlapping joint, shown in Figure 11-9. Both panels have lips which are overlapped during installation, and fasteners are driven through both at once. The fasteners are located in the mortar lines of the stones and are subsequently covered with grout or mortar.

Fig. 11-7. Bathroom with masonry wall.

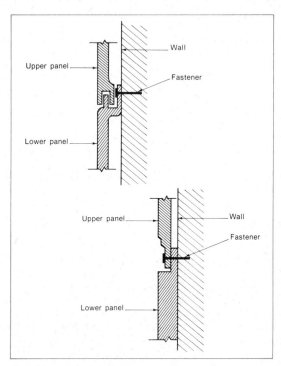

Fig. 11-8. Interlocking panels.

11-2. Tools Needed

Although installation procedures are somewhat different for the three types of panels, the same tools are needed for all:

- Hammer

- Nailset

- Level

- Tape measure or ruler

- Drill with 3/16" bit

- Hacksaw or sabre saw

- Caulking gun

- 3/8" margin trowel.

If the panels are to be installed over existing masonry, the drill bit should have a carbide tip. Any fine-toothed saw may be used, but it should have a carborundum blade to cut through the panels. The caulking gun and trowel are usually supplied with the mortar.

11-3. Installation

Panels may be fastened to the wall by adhesives, nails, or special fasteners. All three types of masonry panels may be installed on solid walls or on horizontal furring. If furring is used, the strips should be spaced 6" between centers. The larger panels of Figures 11-3 and 11-4 may also be fastened directly to open studs. The panels of Figure 11-2 are usually installed beginning at the top and working down to the floor, whereas the other two are installed from the floor level up to the ceiling.

To install this type of panel, begin by placing a level horizontal line 11-1/4" from the ceiling or the top of the installation. If the ceiling is not level, any deviations can later be filled with grout. The first row of panels is installed with its bottom edges along the line.

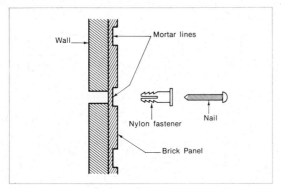

Fig. 11-9. Overlapping joint.

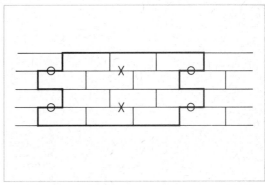

Fig. 11-10. Nylon fasteners.

Start at one corner and work across the wall. If you start at an inside corner, you must saw off the two protruding bricks of the first panel. If you start at an outside corner, you should purchase *outside corners* which are made to cover a corner and interlock with panels on both adjacent walls. If only one wall is to be covered, cut off the two overlapping bricks and fill in the openings. This is done by stuffing paper into the openings and finishing with mortar. When this dries, it may be colored to match the brick with special touch-up material.

Special nylon fasteners are used to hold the panels, as shown in Figure 11-10. A 3/16″ hole is drilled through the mortar line and wall. Then the fastener is inserted in the hole and the nail is driven flush with a nailset, as shown in Figure 11-11. As each panel is installed, two fasteners are inserted to hold it. These

Fig. 11-11. Installing fasteners.

fasteners are positioned as shown by the "X" in Figure 11-10. When the next panel is installed, additional fasteners are inserted in the common junctions as shown by the circles in Figure 11-10. If the wall itself has enough holding power, you can use screws or nails instead of the nylon fasteners. If adhesive is used instead of fasteners, a continuous bead is placed along the back of each mortar line.

Panels in successive rows are brought into light contact with those above them. When you come to the floor, it is unlikely that you will have an exact fit, and the bottom panels will have to be trimmed to fit. Use a saw to cut the bottom panels, or plan to hide the gaps with a base molding. Fitting at windows or around trim also requires cutting. Edges can be filled with mortar.

After the panels are installed, grout or mortar is applied with a caulking gun, as shown in Figure 11-13. The mortar covers the nylon fasteners and also the seams between adjacent panels. The caulking gun is held at a 45 degree angle and slight pressure is applied to make the grout stick to the surface. Always make sure the mortar is at room temperature, since cold mortar is difficult to apply. The mortar bead can be left as it is when expelled from the caulking gun, or can be smoothed with a margin trowel, as shown in Figure 11-14, to achieve whatever mortar shape you prefer. A finished bead without smoothing is shown in Figure 11-15.

The panels shown in Figure 11-3 are installed beginning at the floor. Although not

Fig. 11-13. Apply mortar.

Fig. 11-14. Smoothing mortar.

shown in Figure 11-3, these panels have an interlocking strip that protrudes at each end. This strip must be cut off the edge that fits into a corner. Before beginning, locate the studs and mark their locations on the wall. Panels are nailed directly to the studs with 1-1/2" nails. Usually four nails per panel, along the top edge, are enough.

Make sure panels are level. If the floor is tilted, the panel can be scribed and cut to fit, or a molding can be used later to cover the gap. If the floor is level, a starting strip may be used first at floor level. The first panel locks into the starting strip and is thus held firmly at the bottom. Four nails hold the top firmly. These nails are hidden by the next panel, as shown in Figure 11-8. If no starting strip is used, the lowest panels should be nailed to the studs at the top and bottom. Drill holes in a mortar line near the bottom for the nails. These nailheads can be touched up later.

As each new row of panels is begun, make sure vertical mortar lines are staggered for best visual effect. Each panel is interlocked with the one below, as shown in Figure 11-8. When you get to the top, you will probably have to trim the panel to fit. Then drill holes for nails and nail the top panels to the studs. Touch up the nails later. If a space is left at the top, it can be filled with mortar.

Around windows and doors, you have three options. You can use wood moldings as you would with wood panels. Since the door frames are wooden these moldings would not

look out of place. Stone moldings to match the panels are also available. The third choice is to fill the space with mortar and color it to match the brick. Stone and brick moldings are also available as a cap for wainscotting.

The third type of panel (shown in Figure 11-4) is also installed from the floor up and is nailed directly to studs. Stud locations should be marked on the walls. Start at a corner and make sure the panel is level. Any space at the floor can be covered with a base molding. This is simpler than trimming the panel to fit. As with the second type, mortar lines should be staggered.

When all the panels are in place, mortar is applied to cover all nailheads, and to fill any cracks and openings, and all the recessed mortar joints. The mortar should be smoothed with a margin trowel up to one hour after it is applied. Mortar sets to a waterproof seal in about 2 days.

11-4. Small Openings

Before the panel is installed over a small opening such as an electrical outlet, mark the location on the wall, as shown in Figure 11-16. Then fit the panel to place and mark the panel for cutting around the opening, as shown in Figure 11-17. Cut the panel and nail it in place. Remove the outlet box and reset it

Fig. 11-15. Finished mortar joints.

Fig. 11-17. Marking panel for opening.

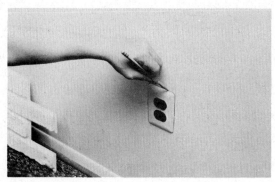
Fig. 11-16. Locating electrical opening.

with longer screws so that the wall plate can be set over the brick or stone.

11-5. Maintenance

Masonry panels need no protective maintenance. Dirt and grease can be removed with water and ordinary household detergents.